# MEI structured mathematics

# Statistics 2

## ANTHONY ECCLES
## NIGEL GREEN
## ROGER PORKESS

### Series Editor: Roger Porkess

**MEI Structure Mathematics** is supported by industry:

BNFL, Casio, GEC, Intercity, JCB, Lucas, The National Grid Company,
Texas Instruments, Thorn EMI

Hodder & Stoughton

A MEMBER OF THE HODDER HEADLINE GROUP

*British Library Cataloguing in Publication Data*

Eccles, Anthony
  Statistics. – Book 2. – (MEI Structured
  Mathematics Series)
  I. Title   II. Series
  519.5

ISBN 0 340 57300 7

First published 1993
Impression number 10 9 8 7 6 5 4
Year 1998 1997 1996 1995

Typeset by Multiplex Techniques Ltd.
Printed in Great Britain for Hodder & Stoughton Educational,
a division of Hodder Headline Plc, 338 Euston Road, London
NW1 3BH by Bath Press, Avon

# MEI Structured Mathematics

Mathematics is not only a beautiful and exciting subject in its own right but also one that underpins many other branches of learning. It is consequently fundamental to the success of a modern economy.

MEI Structured Mathematics is designed to increase substantially the number of people taking the subject post-GCSE, by making it accessible, interesting and relevant to a wide range of students.

It is a credit accumulation scheme based on 45 hour components which may be taken individually or aggregated to give:

3 components     AS Mathematics
6 components     A Level Mathematics
9 components     A Level Mathematics + AS Further Mathematics
12 components  A Level Mathematics + A Level Further Mathematics

Components may alternatively be combined to give other A or AS certifications (in Statistics, for example) or they may be used to obtain credit towards other types of qualification.

The course is examined by the Oxford and Cambridge Schools Examination Board, with examinations held in January and June each year.

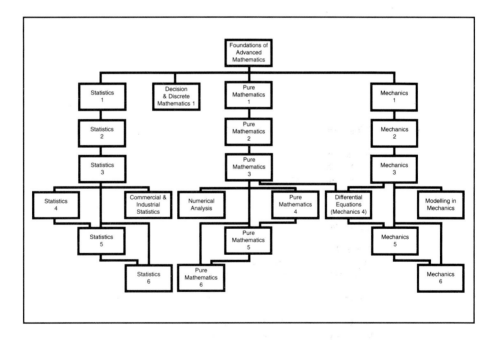

This is one of the series of books written to support the course. Its position within the whole scheme can be seen in the diagram above.

*Mathematics in Education and Industry is a curriculum development body which aims to promote the links between Education and Industry in Mathematics and to produce relevant examination and teaching syllabuses and support material. Since its foundation in the 1960s, MEI has provided syllabuses for GCSE (or O Level), Additional Mathematics and A Level.*

*For more information about MEI Structured Mathematics or other syllabuses and materials, write to MEI Office , 11 Market Street, Bradford-on-Avon BA15 1LL.*

# Introduction

This is the second in a series of books written to support the Statistics Components in MEI Structured Mathematics, but you may also use them for an independent course in the subject. Throughout the course the emphasis is on understanding, interpretation and modelling, rather than on mere routine calculations.

There are four chapters in this book. The techniques for discrete models are covered in the first chapter, and these are illustrated by the Poisson distribution in chapter 2. In chapter 3 you are introduced to the Normal distribution. The book ends with a chapter on bivariate data, covering correlation and regression lines; you meet two correlation coefficients (Pearson's product moment and Spearman's rank) and use them as statistics for suitable hypothesis tests.

Several examples are taken from the pages of a fictional local newspaper, *The Avonford Star*. Much of the information that you receive from the media is of a broadly statistical nature. In these books you are encouraged to recognise this and shown how to evaluate what you are told.

The authors of this book would like to thank the many people who have helped in its preparation and particularly those who read the early versions and trialled them with their students. We would also like to thank the various examining boards who have given permission for their past questions to be included in the exercises.

Anthony Eccles, Nigel Green and Roger Porkess

# Contents

# Discrete Random Variables

*An approximate answer to the right problem is worth a good deal more than an exact answer to an approximate problem.*

John Tukey

THE AVONFORD STAR

# Twins Galore

Out of the 126 babies born in Avonford Maternity Hospital last month, no fewer than 8 were twins, and 3 were triplets.

Hospital superintendent Dr. Fatima Malik said that while unusual there was nothing particularly significant about this; such multiple-birth clusters are for ever occurring somewhere or other she explained.

But self-styled *Guardian of the Environment* Roy James, 55, who lives in Lumumba Drive, says that there must be a source of radioactivity in the area. "It's the logical explanation", he told me.

*Proud Parents Rachel and Stefan Barnes are overjoyed with their new family, but could radioactivity be causing more multiple births like theirs?*

The newspaper article gives two points of view but how do you decide between them?

Before you can make any judgement you must know something about how likely twins, triplets and other multiple births are to occur naturally, that is their *probability distribution*.

| Number of babies | 1 | 2 | 3 | 4+ |
|---|---|---|---|---|
| Probability | 0.989 | 0.010 | 0.001 | Negligible |

How does this help you to evaluate the two claims?

There have been 4 pairs of twins and one set of triplets from 120 pregnancies, giving 126 babies in all.

The situation can be modelled using the binomial distribution. Starting with the triplets, the probability of 1 set of triplets is

$$^{120}C_1 0.999^{119}0.001^1 = 0.107.$$

For the remaining 119 mothers, the conditional probabilities of singles and twins, given that they have not had triplets, are $\dfrac{0.989}{0.999} \approx 0.99$ and $\dfrac{0.010}{0.999} \approx 0.01$

Thus the probability of 4 pairs of twins is

$$^{119}C_4 0.99^{115}0.01^4 = 0.025.$$

So the combined probability is $0.107 \times 0.025 = 0.0027$ or about $1/375$.

So you would expect that particular outcome about one month in every 30 years in a hospital of Avonford's size. Given the number of such hospitals up and down the country, the doctor's comment that such multiple-birth clusters are always occurring somewhere or other seems perfectly reasonable. If, however, Avonford continues to have high rates of multiple births in the months ahead, then Roy James's claim should be investigated along with other possible causes.

## Statistical Models

You will find people who will tell you that you can extend the pattern for the probabilities of multiple births beyond triplets:

| Number of babies | 2 | 3 | 4 | 5 | ... |
|---|---|---|---|---|---|
| Probability | 0.010 | 0.001 | 0.0001 | 0.00001 | ... |

so that the probability for $n$ babies at one pregnancy is given by

$$P = (0.1)^n \qquad \text{for } n > 1$$

$$\text{and} \quad P = 1 - \sum_{r=2}^{\infty}(0.1)^n = 0.988888 \qquad \text{for } n = 1.$$

This is a *mathematical model* to describe the situation. Unfortunately, although algebraically neat, it is a very poor model. The probability of quintuplets is more like one in a hundred million than the one in a hundred thousand suggested by this model.

In Statistics you are often looking for models to describe and explain the data you find in the real world. In this chapter you are introduced to some of the techniques for working with models for discrete data, but you should always remember that your results can never be better than your model.

It is always tempting to look for tidy algebraic relationships, like the one proposed for multiple births, but in practice this may just not be possible.

In some cases, like tossing an unbiased coin or rolling a fair die, the model is completely accurate, but in others this is not so. In Chapters 2 and 3 of this book you meet two standard distributions, the Poisson and the normal. These, and other theoretical distributions, are often used to model real life situations.

The fit can be very good, but you should not forget that they are only being used as models to describe the actual situation. A model which is based on data, like that for the probability of multiple births, can never be more accurate than the actual data, so you should always be prepared to investigate the source of the figures.

The probabilities given for multiple births are derived from national data collected over many years.

# Discrete Random Variables

The number of babies at any pregnancy is an example of a *discrete random variable*.

It is *discrete* because it takes only particular values, 1, 2, 3, 4 etc; you cannot have 2.4 or 0.315 babies. By contrast a baby's birth weight is a *continuous variable* which can take any value between reasonable limits.

It is *random* because on becoming pregnant a woman cannot determine how many babies she is going to have. Having twins or triplets is a chance event. (The effects of fertility drugs and in vitro fertilisation have been ignored.)

It is a *variable* because it takes different numerical values.

## Notation

A random variable is usually denoted by an upper case letter, such as $X$, $Y$, or $Z$, etc. You may think of this as the name of the variable. The particular values the variable takes are denoted by lower case letters, such as $x$, $y$, $z$. Sometimes these are given suffices $x_1, x_2, x_3$ etc.

# The conditions for a Discrete Random Variable

If the outcome of a process can be stated as a number, $X$, which can take different possible values $x_1, x_2, \ldots, x_n$ at random, then $X$ is a discrete random variable. The probabilities $p_1, p_2, \ldots, p_n$ of the different values must sum to 1.

| Outcome | $x_1$ | $x_2$ | $x_3$ | $\ldots$ | $x_n$ |
|---|---|---|---|---|---|
| Probability | $p_1$ | $p_2$ | $p_3$ | $\ldots$ | $p_n$ |

$$p_1 + p_2 + p_3 + \ldots + p_n = 1 \qquad p_i \geq 0, \text{ for } i = 1, \ldots, n.$$

Another way of saying this is that the various outcomes cover all possibilities, that is they are *exhaustive*.

**N O T E**   *Since $p_1$, $p_2$ etc are probabilities, none of them may exceed 1.*

*This gives the probability distribution of X. The rule which assigns probabilities to the various outcomes is known as the probability function of X. Sometimes it is possible to write the probability distribution as a formula.*

# Diagrams of probability distributions

If you wish to draw a diagram to show the distribution of a discrete random variable, you must show clearly that the variable is indeed discrete, as in the two diagrams of figure 1.1.

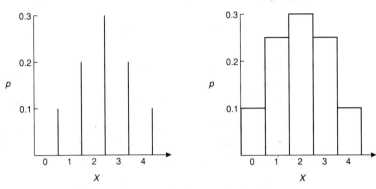

**Figure 1.1**

This may be contrasted with the distribution of a *continuous* variable which is drawn as a continuous curve, figure 1.2. Its equation is called the *probability density function* and usually written f(x). The area under this curve represents probability.

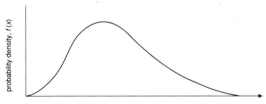

**Figure 1.2**

You will often find yourself wanting to know the probability that a random variable takes a value no more than a certain number, P(X≤x). This is known as the *cumulative distribution function* and generally denoted by F(x), see Figure 1.3.

(i) F(1)                                                          (ii) F(a)

**Figure 1.3 The shaded area is the Cumulative Distribution Function.**

**EXAMPLE**

The variable $X$ is the score on an unbiased die after it is rolled.
(i)   Write down the probability distribution of $X$.
(ii)  Show that $X$ satisfies the conditions for it to be a discrete random

*Solution:*

(i)   The probability distribution of $X$ is:

| Outcome | 1 | 2 | 3 | 4 | 5 | 6 |
|---|---|---|---|---|---|---|
| **Probability** | $1/6$ | $1/6$ | $1/6$ | $1/6$ | $1/6$ | $1/6$ |

Notice that $P(X=1) + P(X=2) + \ldots + P(X=6) = 1$,

because $1/6 + 1/6 + \ldots + 1/6 = 1$.

(ii)  The set of possible values of $X$, $\{1,2,3,4,5,6\}$ form a discrete set. The outcome from rolling a die is clearly random.

Therefore $X$ is discrete random variable.

---

**EXAMPLE**

A card is selected at random from a normal pack of 52 playing cards. If it is a Spade the experiment ends, otherwise it is replaced, the pack is shuffled and another card is selected. What is the probability distribution of $X$, the number of cards selected up to and including the first spade?

*Solution:*

This situation is shown in the tree diagram below. You will see that there is no limit to the value which $X$ may take, although larger values of $X$ become more and more unlikely.

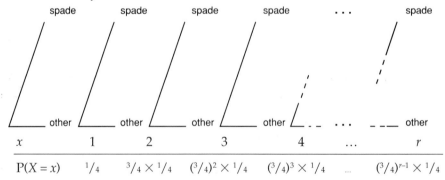

| $x$ | 1 | 2 | 3 | 4 | ... | $r$ |
|---|---|---|---|---|---|---|
| $P(X = x)$ | $1/4$ | $3/4 \times 1/4$ | $(3/4)^2 \times 1/4$ | $(3/4)^3 \times 1/4$ | ... | $(3/4)^{r-1} \times 1/4$ |

So the probability distribution is given by

$P(X=r) = (3/4)^{r-1} \times 1/4 \qquad$ for $r = 1,2,3\ldots$

*Check:*

Since this is a probability distribution, the sum of all the probabilities should be 1. Is it true that

$$1/4 + 3/4 \times 1/4 + (3/4)^2 \times 1/4 + (3/4)^3 \times 1/4 + \ldots + (3/4)^{r-1} \times 1/4 + \ldots = 1?$$

You will recognise that this is an infinite geometric series. The first term, $a = \frac{1}{4}$, and the common ratio, $r = \frac{3}{4}$.

Substituting these values in the formula for the sum of an infinite geometric series gives:

$$S_\infty = \frac{a}{1-r} = \frac{\frac{1}{4}}{1-\frac{3}{4}} = \frac{\frac{1}{4}}{\frac{1}{4}} = 1 \text{ as required.}$$

**EXAMPLE**

The probability distribution of a random variable $Y$ is given by

$$P(Y = y) = cy \qquad \text{for } y = 1,2,3,4,5.$$

(i)   Given that $c$ is a constant find the value of $c$.
(ii)  Hence find the probability that $Y > 3$.

*Solution:*

(i)   The relationship $P(Y = y) = cy$ gives the table

| $Y$ | 1 | 2 | 3 | 4 | 5 |
|------|------|------|------|------|------|
| $P(Y)$ | $c$ | $2c$ | $3c$ | $4c$ | $5c$ |

and since $Y$ is a random variable $c + 2c + 3c + 4c + 5c = 1$

$$15c = 1$$
$$c = \frac{1}{15}.$$

(ii)  $P(Y>3) = P(Y=4) + P(Y=5)$

$$= \frac{4}{15} + \frac{5}{15}$$
$$= \frac{9}{15} = \frac{3}{5}$$

**N O T E**

*You may have found the expression $P(Y= y)$ somewhat curious. You should read it as 'The probability, P, that the random variable called Y has the particular value y'.*

## Exercise 1A

**1**. The probability distribution of a discrete random variable, $X$, is given by

$$P(X=x) = kx$$

where $k$ is a constant, for $x = 1, 2, 3, 4$.

Find the value of $k$.

**2**. The probability that a variable, $Y$, takes the value $y$ is given by

$$P(Y=y) = (\tfrac{1}{6})(\tfrac{5}{6})^{y-1} \quad \text{for } y = 1, 2, 3 \ldots$$

Show that $Y$ satisfies the conditions for it to be a discrete random variable and suggest a situation it could be modelling.

3. The probability distribution of a discrete random variable $Z$ is given by

$$P(Z = z) = az/8 \quad \text{for } z = 2,4,6,8.$$

   (i) Find the value of $a$.
   (ii) Hence find $P(Z < 6)$.

4. The random variable $X$ is given by the number of heads obtained when 5 fair coins are tossed. Write out the probability distribution for $X$.

5. The probability distribution of a discrete random variable $Y$ is given by

$$P(Y = y) = cy/5 \quad \text{for } y = 1,2,3,4,5,$$

   where $c$ is a constant.

   Find the value of $c$ and the probability that $Y$ is less than 3.

6. The random variable $X$ is given by the sum of the scores when two ordinary dice are thrown.

   Write out the probability distribution of $X$ and verify that $X$ is a discrete random variable.

7. The random variable $Y$ is the difference of the scores when two ordinary dice are thrown.

   (i) Write out the probability distribution of $Y$.
   (ii) Find $P(Y < 3)$.

8. The random variable $Z$ is the number of heads obtained when four fair coins are tossed.

   (i) Write out the probability distribution of $Z$.
   (ii) Find the probability that there are more heads obtained than tails.

9. A box contains six black and four red pens. Three pens are taken from the box. The random variable $X$ is the number of red pens obtained.

   Find the probability distribution of $X$.

10. Three committee members are to be selected from 6 men and 7 women. Write out the probability distribution for the number of men appointed to the committee assuming the selection is done at random.

11. Two tetrahedral dice each with faces labelled 1, 2, 3 and 4 respectively thrown and the random variable $X$ is the product of the numbers on which the dice fall.

    (i) Find the probability distribution of $X$.

    (ii) What is the probability that any throw of the dice results in a value of $X$ which is an odd number?

12. Four cards are drawn, without replacement from a normal pack of 52. Write the probability distribution for the number of red cards chosen.

**13.** An ornithologist carries out a study of the numbers of eggs laid per pair by a species of rare bird in its annual breeding season. He concludes that it may be considered as a discrete random variable $X$ with probability distribution given by

$P(X = 0) = 0.2$

$P(X = x_i) = k(4x_i - x_i^2)$ for $x_i = 1, 2, 3$ or $4$

$P(X = x_i) = 0$ for $x_i > 4$.

(i) Find the value of $k$ and write out the probability distribution as a table. The ornithologist observes that the probability of survival (that is of an egg hatching and of the chick living to the stage of leaving the nest) is dependent on the number of eggs in the nest. He estimates these probabilities to be as follows.

| $x_i$ | Probability of survival |
|-------|-------------------------|
| 1 | 0.8 |
| 2 | 0.6 |
| 3 | 0.4 |

(ii) Find, in the form of a table, the probability distribution of the number of chicks surviving per pair of adults.

**14.** A sociologist is investigating the changing pattern of the numbers of children which women have in a country. She denotes the present number by the random variable $X$ which she finds to have the following distribution.

| Number of children $x_i$ | 0 | 1 | 2 | 3 | 4 | 5+ |
|---------------------------|------|------|-----|------|------|------------|
| Probability $P(X = x_i)$ | 0.09 | 0.22 | $a$ | 0.19 | 0.08 | negligible |

(i) Find the value of $a$.

She is anxious to find an algebraic expression for the probability distribution and tries

$P(X = x_i) = k(x_i + 1)(5 - x_i)$ for $0 \le x_i \le 5$, ($x_i$ may only take integer values)

$= 0$ otherwise

(ii) Find the value of $k$ for this model.

(ii) Compare the algebraic model with the probabilities she found. Do you think it is a good model? Is there any reason that why it must be possible to express the probability distribution in a neat algebraic form?

**15.** In a game, each player throws four ordinary six-sided dice. The random variable $X$ is the largest number showing on the dice.

(i) Find the probability that $X = 1$.

(ii) Find the probability that $X \le 2$ and deduce that the probability $X = 2$ is $^5/_{432}$.

MEI Structured Mathematics

*Exercise 1a continued*

(iii) Find the probability that $X = 3$.

(iv) Find the probability that $X = 6$, and explain without further calculation why 6 is the most likely value of $X$.

[MEI]

# Expectation

This advertisement was placed by Claire Dadzie who designs and prints tee-shirts. She sells them direct to the public for £7 each, avoiding all middlemen.

Claire's tee-shirts come in four sizes: small, medium, large and extra large and her profits per sale are £4, £3.50, £3 and £2.50 respectively. She believes that 30% of her orders are for small size, 40% for medium size, 20% for large size and 10% for extra large.

How much profit can she expect to make if she sells 500 tee-shirts?

What is her average profit per sale?

Such questions, which are very important to Claire if she is to stay in business, involve the idea of *expectation*.

Claire has a model for the probability distribution of different values of her profit per sale:

| Size | Small | Medium | Large | Extra Large |
|---|---|---|---|---|
| **Profit (£)** | 4.00 | 3.50 | 3.00 | 2.50 |
| **Probability** | 0.3 | 0.4 | 0.2 | 0.1 |

If she sells 500 tee-shirts, she expects their sizes to be as follows:

Small         $500 \times 0.3 = 150$
Medium        $500 \times 0.4 = 200$
Large         $500 \times 0.2 = 100$
Extra Large   $500 \times 0.1 = \underline{\phantom{0}50}$
              Total          $500$

In this case her total profit will be

$$\text{Profit} = 150 \times £4.00 \ + \ 200 \times £3.50 \ + \ 100 \times £3.00 \ + \ 50 \times £2.50$$
$$= £1725$$

Her expected average profit per sale will be $£1725 \div 500 = £3.45$.

You could have found these answers in the reverse order, as follows:

Her expected profit per sale is given by:

$$0.3 \times £4.00 \ + \ 0.4 \times £3.50 \ + \ 0.2 \times £3.00 \ + \ 0.1 \times £2.50 \ = \ £3.45.$$

Her expected profit on 500 sales is given by:

$$500 \times £3.45 = £1725.$$

You will see at once that the two approaches are equivalent to each other. The first involved working out the expected profit for a large number of tee-shirts and using that to deduce the profit for one; the second started by finding the profit on one tee-shirt and multiplying up to find that on a large number of them. The first step of the second approach may easily be generalised to define expectation.

*Definition*:

If a discrete random variable, $X$, takes possible values $x_1, x_2, x_3, \ldots$ with associated probabilities $p_1, p_2, p_3, \ldots$, the expectation $\text{E}(X)$ of $X$ is given by:

$$\text{E}(X) = \sum_i x_i p_i.$$

Thus in the case of Claire's tee-shirts,

$$\text{E(Profit)} = £4.00 \times 0.3 \ + \ £3.50 \times 0.4 \ + \ £3.00 \times 0.4 \ + \ £2.50 \times 0.1 \ = \ £3.45$$
$$\phantom{\text{E(Profit)} =} x_1 p_1 \quad + \quad x_2 p_2 \quad + \quad x_3 p_3 \quad + \quad x_4 p_4$$

**N O T E**   *The terms* average *and* mean *are often used to convey the same idea as expectation but this may cause confusion.*

*Expectation is actually the mean of the underlying distribution, the* parent population. *Expectation is denoted by* $\mu$ *pronounced 'mew'). The terms average and mean can be applied to either the parent population or to a particular sample.*

*So if one day Claire sells 200 tee-shirts at an overall profit of £600, that is a sample of her overall sales over a much longer period. On that one day her average or mean profit is £600 ÷ 200 = £3.00.*

*The expectation of her profit however is her mean long-term profit per tee-shirt, and is £3.45.*

*Figures defining the parent population are called* parameters *and usually denoted by Greek letters. Thus the letter $\mu$ is used for expectation.*

**EXAMPLE**

What is the expectation score when an unbiased die is rolled once?

*Solution*

The probability distribution is:

| Outcome | 1 | 2 | 3 | 4 | 5 | 6 |
|---|---|---|---|---|---|---|
| **Probability** | $1/6$ | $1/6$ | $1/6$ | $1/6$ | $1/6$ | $1/6$ |

The expectation, $E(X) = \Sigma x_i p_i$

$$= 1 \times {}^1/_6 + 2 \times {}^1/_6 + 3 \times {}^1/_6 + 4 \times {}^1/_6 + 5 \times {}^1/_6 + 6 \times {}^1/_6$$

$$= {}^{21}/_6 = 3.5.$$

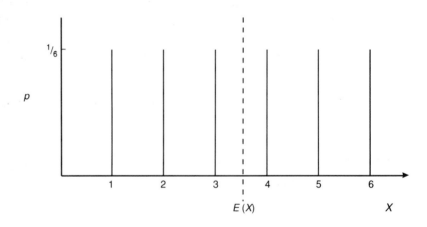

Notice that in this case the value of 3.5 for $E(X)$ is an impossible outcome. That is quite often the case (eg an average family with 2.4 children) and it is wrong to round it to the nearest apparently sensible value. When used in Statistics, expectation is a technical term, with a different meaning from that in everyday English.

**EXAMPLE**

A fruit machine is constructed so that the cost is 10p per turn, and the amount won by the user in pence at each turn has the following probability distribution:

| Payout (p) | Nil | 50 | 100 | 500 |
|---|---|---|---|---|
| Less stake | −10 | −10 | −10 | −10 |
| Winnings | −10 | 40 | 90 | 490 |
| Probability | 0.94 | 0.03 | 0.02 | 0.01 |

(i) What is the expected loss per turn?

(ii) How much can the user expect to lose after 20 turns?

*Solution*

Let the random variable $X$ be the amount won per turn in pence by the user.

(i) $E(X) = -10 \times 0.94 + 40 \times 0.03 + 90 \times 0.02 + 490 \times 0.01$

$= -9.4 + 7.9$

$= -1.5\text{p}$

The expectation of the amount won per turn is −1.5 pence, ie a loss of 1.5 pence.

(ii) After 20 turns the user can expect to lose $20 \times 1.5\text{p} = 30\text{p}$.

# Expectation of a function of $X$, $E(g[X])$

Sometimes you will need to find the expectation of a function of a random variable. That sounds rather forbidding and you may think the same of the definition given below at first sight. However, as you will see in the next two examples, the procedure is straightforward and common sense.

*Definition*

If $g[X]$ is a function of the discrete random variable $X$ then $E(g[X])$ is given by

$$E(g[X]) = \sum_i g[x_i]P(X = x_i)$$

**EXAMPLE** What is the expectation of the square of the number that comes up when a fair die is rolled?

*Solution*

Let the random variable $X$ be the number that comes up when the die is rolled.

$$g[X] = X^2$$

$$E(g[X]) = E(X^2) = \sum_i x_i^2 P(X = x_i)$$

$$= 1^2 \times {}^1/_6 + 2^2 \times {}^1/_6 + 3^2 \times {}^1/_6 + 4^2 \times {}^1/_6 + 5^2 \times {}^1/_6 + 6^2 \times {}^1/_6$$

$$= 1 \times {}^1/_6 + 4 \times {}^1/_6 + 9 \times {}^1/_6 + 16 \times {}^1/_6 + 25 \times {}^1/_6 + 36 \times {}^1/_6$$

$$= {}^{91}/_6$$

$$= 15.17.$$

**NOTES**

1. $E(X^2)$ is not the same as $[E(X)]^2$. In this case $15.17 \neq 3.5^2$ which is $12.25$.

2. This calculation could also have been set out in table form as shown below.

| $x_i$ | $P(X=x_i)$ | $x_i^2$ | $x_i^2 P(X=x_i)$ |
|-------|------------|---------|------------------|
| 1 | $^1/_6$ | 1 | $^1/_6$ |
| 2 | $^1/_6$ | 4 | $^4/_6$ |
| 3 | $^1/_6$ | 9 | $^9/_6$ |
| 4 | $^1/_6$ | 16 | $^{16}/_6$ |
| 5 | $^1/_6$ | 25 | $^{25}/_6$ |
| 6 | $^1/_6$ | 36 | $^{36}/_6$ |
| | | | $\Sigma \quad ^{91}/_6$ |

$$E(g[X]) = {}^{91}/_6 = 15.17$$

**EXAMPLE** A random variable $X$ has the following probability distribution:

| Outcome | 1 | 2 | 3 |
|---------|-----|-----|-----|
| Probability | 0.4 | 0.4 | 0.2 |

(i) Calculate $E(4X + 5)$
(ii) Calculate $4E(X) + 5$
(iii) Comment on the relationship between your answers to parts (i) and (ii).

*Solution*

(i) To find $E(4X + 5)$

Using $E(g[X]) = \Sigma\, g[x_i] \cdot P(X=x_i)$ with $g[X] = 4X + 5$,

| $x_i$ | 1 | 2 | 3 |
|-------|-----|------|------|
| $g(x_i)$ | 9 | 13 | 17 |
| $P(X=x_i)$ | 0.4 | 0.4 | 0.2 |

$$E(4X + 5) = E[g(X)]$$
$$= 9 \times 0.4 + 13 \times 0.4 + 17 \times 0.2$$
$$= 12.2.$$

(ii) To find $4E(X) + 5$

$$E(X) = 1 \times 0.4 + 2 \times 0.4 + 3 \times 0.2 = 1.8$$

and so

$$4E(X) + 5 = 4 \times 1.8 + 5$$
$$= 12.2.$$

(iii) Clearly $E(4X + 5) = 4E(X) + 5$, both having the value 12.2.

# Expectation Algebra

In the example above you found that E(4X + 5) = 4E(X) + 5.

The working was numerical, showing that both expressions came out to be 12.2, but it could also have been shown algebraically. This would have been set out as follows:

| Proof | Reasons (general rules) |
|---|---|
| $E(4X + 5) = E(4X) + E(5)$ | $E(X \pm Y) = E(X) \pm E(Y)$ |
| $= 4E(X) + E(5)$ | $E(aX) = aE(X)$ |
| $= 4E(X) + 5$ | $E(c) = c$ |

Look at the general rules on the right hand side of the page. ($X$ and $Y$ are random variables, $a$ and $c$ are constants). They are important but they are also common sense.

Notice the last one, which in this case means the expectation of 5 is 5. Of course it is; 5 cannot be anything else but 5. It is so obvious that sometimes people find it confusing!

These rules can be extended to take in the expectation of the sum of two functions of a random variable.

$E(f[X] + g[X]) = E(f[X]) + E(g[X])$   where f and g are both functions of $X$.

*Proof*

By definition

$$E(f[X] + g[X]) = \sum_i (f[x_i] \qquad + \quad g[x_i])P(X = x_i)$$

$$= \sum_i f[x_i] \cdot P(X = x_i) \quad + \quad \sum_i g[x_i] \cdot P(X = x_i)$$

$$= E(f[X]) \qquad\qquad + \quad E(g[X]) \qquad\qquad \text{as required}$$

## Exercise 1B

1. Find the expectation of the number of tails when three fair coins are tossed.

2. Find the expectation for the outcome with the following distribution:

| Outcome | 1 | 2 | 3 | 4 | 5 |
|---|---|---|---|---|---|
| Probability | 0.2 | 0.2 | 0.4 | 0.1 | 0.1 |

3. A discrete random variable $X$ can assume only the values 4 and 5, and has expectation 4.2. Find the two probabilities $P(X = 4)$ and $P(X = 5)$.

4. A discrete random variable $Y$ can take only the values 50 and 100. Given that $E(Y) = 80$, write out the probability distribution of $Y$.

**5.** The probability distribution of a discrete random variable $Z$ is given by

$$P(Z = z) = cz \quad \text{for } z = 2,3,4$$

where $c$ is a constant. Find $E(Z)$.

**6.** A man buys eight tickets from a total of 2000 tickets in a raffle where there is just one prize of £50. The price of a ticket is 10p. Given that all the tickets are sold, calculate his expected loss.     [L]

**7.** An unbiased tetrahedral die has faces labelled 2, 4, 6 and 8. If the die lands on the face marked 2, the player has to pay 50p. If it lands on a face marked with a 4 or a 6 the player wins 20p, and if it lands on the face labelled 8 then no money changes hands.

Find the expected gain or loss of the player after

(i) 1 throw; (ii) 3 throws; (iii) 100 throws.

**8.** Melissa is planning to start a business selling ice-cream. She presents a business plan to her bank manager in which she models the situation as follows.

'Three tenths of days are *warm*, two fifths are *normal* and the rest are *cool*. On warm days I will make £100 (that is the difference between what I get from selling ice cream and the cost of the ingredients), on normal days £70 and on cool days £50. My total daily costs, will be £45.'

(i)  Find Melissa's expected profit per six day week.
(ii) If you were her bank manager, how much money would you be prepared to lend her?

**9.** The discrete random variable $X$ has probability distribution given by

$$P(X = x) = \frac{(4x + 7)}{68} \quad \text{for } x = 1, 2, 3, 4.$$

Find (i) $E(X)$; (ii) $E(X^2)$; (iii) $E(X^2 + 5X - 2)$; (iv) verify that $E(X^2 + 5X - 2) = E(X^2) + 5E(X) - 2$.

**10.** A discrete random variable $X$ has a probability distribution as follows:

| $x$ | 0 | 1 | 5 | 20 | $w$ |
|---|---|---|---|---|---|
| $P(X = x)$ | 0.1 | 0.2 | 0.3 | 0.1 | 0.3 |

Find the value of $w$ in the two cases

(i) $E(X) = 10$; (ii) $E(X) = 12$.

**11.** In a discrete probability distribution, the variable $X$ takes the value $120/k$ with probability $k/45$, where $k$ takes all positive integral values from 1 to $n$ inclusive.

Verify that $n = 9$.

(ii) Calculate the probability that $X$ takes the value 20.

(iii) Find the probability that the value of $X$ lies between 14 and 41.

(iv) Calculate the expected value of $X$. [L]

12. A lady has 5 coins in a box: two £1, two 20p and one 10p. She wants the 10p coin to use in a slot machine, and takes the coins out at random, one at a time until she comes to the one she wants. She does not replace a coin once she has taken it out of her pocket. Find

(i) the expectation of the number of coins she takes out;

(ii) the expectation of the amount of money she takes out.

(In both cases include the final 10p).

13. An experiment consists of throwing two unbiased dice and recording $r$, the higher of the two numbers shown. When each die shows the same number, this is taken as the value of $r$. Complete the entries in the following tables, where $p_1, p_2, \ldots$ are the probabilities that $r = 1, 2 \ldots$.

| $p_1$ | $p_2$ | $p_3$ | $p_4$ | $p_5$ | $p_6$ |
|-------|-------|-------|-------|-------|-------|
| $^1/_{36}$ | $^3/_{36}$ | | | | |

Verify that the mean of $r$ is $^{161}/_{36}$.

The experiment, as described above, is done *three* times. Find the probabilities that

(i) the three values of $r$ are $3, 4, 5$ in any order;

(ii) the sum of the three values of $r$ is 16. [MEI]

14. A radio network is launching a new music game (based on one actually broadcast in France). Each contestant is given the title of a song and a list of 7 words, exactly 3 of which occur in the lyric of the song. The contestant is asked to choose the 3 correct words (the choice being made before listening to the song). Assuming that a particular contestant makes this choice completely by guesswork, find the probabilities that he gets $0, 1, 2, 3$ words correct.

Prizes of £1, £3 and £$r$ are to be awarded for $1, 2$ and $3$ words correct respectively. Find, in terms of $r$, the expected prize paid to a contestant choosing completely by guesswork. Hence determine the greatest integer value of $r$ for which the expected prize is less than £3. [MEI]

15. A company takes on new employees in batches of 80 and immediately gives them training in how to operate and maintain a computer system.

The training course lasts one week, at the end of which the trainees are given a test. They are allowed to take the test up to three times but a trainee who has passed the test does not take it again. Those who fail three times are not employed.

In one batch of 80 trainees, the results were as follows:

| | Start | Pass | Fail |
|---|---|---|---|
| **Week** 1 | 80 | 32 | 48 |
| **Week** 2 | 48 | 36 | 12 |
| **Week** 3 | 12 | 3 | 9 |

The company believes this batch to be typical and decides to use these figures to model the costing of its training programme.

(i) Find the probability that someone passes who is attempting the test at: (A) the first time; (B) the second time; (C) the third time.

(ii) Use your answers to part (i) to deduce the probability that someone fails all 3 times, and show that the same answer can be obtained by looking at the overall figures.

(iii) Find the probability that someone who passed the test did so at the second attempt.

It costs the company £4000 overheads + £200 per person for each week that the course is run; (iv) find the expected cost per successful trainee. The company considers, as alternative policies, running the course for only one or two weeks; (v) which of the three policies (1, 2 or 3 weeks) would result in the lowest expected cost per trained operator?

# Variance

The standard deviation of a discrete random variable gives you a measure of the spread of the distribution about the mean or expected value. The variance is simply the square of the standard deviation. The definition of variance of a discrete random variable is very similar to that used when finding the variance of a set of numbers.

*Definition:*

The variance of a discrete random variable $X$, Var($X$), is given by

$$\text{Var}(X) = E([X - \mu]^2).$$

Another, and often more convenient form of this definition is

$$\text{Var}(X) = E(X^2) - \mu^2.$$

*Proof that the two forms are equivalent:*

$$\text{Var}(X) = E([X - \mu]^2)$$
$$= E(X^2 - 2\mu X + \mu^2)$$
$$= E(X^2) - 2\mu E(X) + E(\mu^2)$$
$$= E(X^2) - 2\mu^2 + \mu^2$$
$$= E(X^2) - \mu^2.$$

This result is sometimes written in the form

$$\begin{aligned} \text{Var}(X) = E(X^2) &\quad - \quad [E(X)]^2 \\ = \text{expectation} &\quad - \quad \text{the square of} \\ \text{of the square} &\quad \quad \text{the expectation} \end{aligned}$$

THE AVONFORD STAR

# Consumer Concern

**Mel Charles, Consumer Affairs correspondent.**

Dear Mel,

I am an old age pensioner and my main hobby is gardening. Recently I bought this packet of very expensive seeds from a local company. There were 8 seeds in the packet and it said:

*"The average germination rate for these seeds is about 50%".*

I followed the planting instructions exactly but not one of them came up. So I went out and bought another packet but exactly the same thing happened. I really want to grow these plants but I certainly can't afford to go on buying dud seeds.

Can you help me?

Yours sincerely,

Reg Harper

(Empty seed packet enclosed).

This is just one of many letters I have received about these seeds which are being sold by The Avonford Seed Co. I went to see their Managing Director, Jill Yates, who was genuinely concerned at the problem and brought in their Research Officer, James Wilkinson, to investigate what had gone wrong.

James set up an experiment. He planted the seeds from a large number of packets and counted the number germinating from each packet. The results surprised him. Given in percentage terms they are:

| Number of seeds germinating | 0 | 1 | 2 | 3 | 4 | 5 | 6 | 7 | 8 |
|---|---|---|---|---|---|---|---|---|---|
| % of packets | 36 | 8 | 4 | 0 | 0 | 1 | 3 | 13 | 35 |

You can easily work out the average germination rate for the seeds from these figures; it is 51.25%. The claim on the packet was true but it certainly did not tell the whole story.

What is the whole story?

It is immediately obvious from the figures that the distribution is very spread out, with most packets having either 0 or 8 seeds germinating. Not one packet had 50% of its seeds germinating.

Clearly you also need to know something about the spread of these figures.

The variance is calculated using $\text{Var}(X) = E(X^2) - \mu^2$. The random variable $X$ is the number germinating and the percentages found in the experiment are written as probabilities. Since it was a large scale experiment they should be reasonably accurate.

| $x$ | 0 | 1 | 2 | 3 | 4 | 5 | 6 | 7 | 8 |
|---|---|---|---|---|---|---|---|---|---|
| $P(X = x)$ | 0.36 | 0.08 | 0.04 | 0 | 0 | 0.01 | 0.03 | 0.13 | 0.35 |
| $x^2$ | 0 | 1 | 4 | 9 | 16 | 25 | 36 | 49 | 64 |

The calculation is then as follows:

$$E(X) = \mu = 0 \times 0.36 + 1 \times 0.08 + 2 \times 0.04 + 3 \times 0 + 4 \times 0 + 5 \times 0.01 + 6 \times 0.03$$
$$+ 7 \times 0.13 + 8 \times 0.35$$
$$= 4.1$$

$$\text{Var}(X) = [0 \times 0.36 + 1 \times 0.08 + 4 \times 0.04 + 9 \times 0 + 16 \times 0 + 25 \times 0.01$$
$$+ 36 \times 0.03 + 49 \times 0.13 + 64 \times 0.35] - 4.1^2$$
$$= 13.53.$$

So the standard deviation is given by $\sqrt{13.53} = 3.7$ (to one decimal place).

This tells you that a typical value of $X$ is about 3.7 above or below the mean, that is at one extreme of the range or the other. If you knew this you would be warned to expect the distribution to be bidmodal or U-shaped, see Figure 1.4.

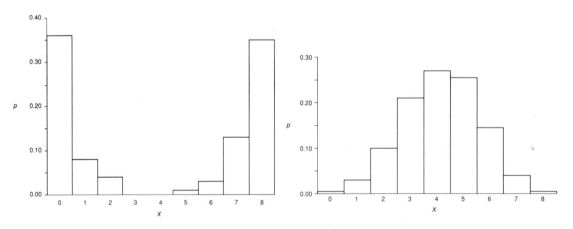

**Figure 1.4 Probability distribution of $X$, the number of seeds germinating from a packet of 8 seeds**     **Figure 1.5 Probability distribution of B(8,0.5125)**

You would actually expect this to be a binomial distribution, B(8,0.5125), see Figure 1.5.

So what has gone wrong? The article went on:

When I saw these results I was perplexed but James Wilkinson knew at once what had happened. "We import these seeds prepacked," he said. "They are a cottage industry in a village in the foothills of the Himalayas. We supply the packets and the villagers put them in. Clearly the seeds from some plants are fertile and those from others not. They don't know which is which but most packets must have all their seeds from the same plant. The solution is easy. All we needs to do is to unpack all the seeds, mix them up thoroughly and repack them."

And Jill Yates has asked me to send a replacement packet to everyone who wrote in complaining. "We are really grateful for your help in sorting out this problem," she told me.

So why was the distribution not binomial?

The answer is that the trials, in this case whether a seed germinated or not, were not independent. If one seed in a packet did not germinate, it was more likely that the same would be true for the others.

**EXAMPLE** The discrete random variable $X$ has probability distribution as follows:

| $x_i$ | 0 | 1 | 2 | 3 |
|-------|-----|-----|-----|-----|
| $P(X = x_i)$ | 0.2 | 0.3 | 0.4 | 0.1 |

Find (i) $E(X)$;  (ii) $E(X^2)$;  (iii) $Var(X)$ using (A) $E(X^2) - \mu^2$; (B) $E([X -\mu]^2)$.

*Solution*

(i) $E(X)$ $= \Sigma x_i P(X = x_i)$

$= 0 \times 0.2 \ + \ 1 \times 0.3 \ + \ 2 \times 0.4 \ + \ 3 \times 0.1$

$= 1.4.$

(ii) $E(X^2)$ $= \Sigma x_i^2 P(X = x_i)$

$= 0 \times 0.2 \ + \ 1 \times 0.3 \ + \ 4 \times 0.4 \ + \ 9 \times 0.1$

$= 2.8.$

(iii) (A) $Var(X) = E(X^2) - \mu^2$

$= 2.8 - 1.4^2$

$= 0.84.$

(B) $Var(X) = E([X-\mu]^2)$

$= \Sigma (x_i-\mu)^2 P(X=x_i)$

$= (0-1.4)^2 \times 0.2 \ + \ (1-1.4)^2 \times 0.3 \ + \ (2-1.4)^2 \times 0.4 \ +$
$(3-1.4)^2 \times 0.1$

$= 0.392 \ + \ 0.048 \ + \ 0.144 \ + \ 0.256$

$= 0.84.$

Notice that the two methods of calculating the variance in part (iii) give the same result, as of course they must.

---

**EXAMPLE** The random variable $X$ has the following probability distribution:

| $x$ | 1 | 2 | 3 | 4 |
|---|---|---|---|---|
| $P(X = x)$ | 0.6 | 0.2 | 0.1 | 0.1 |

Find (i) Var($X$); (ii) Var(7); (iii) Var($3X$); (iv) Var($3X + 7$).

What general results do answers (ii) to (iv) illustrate?

*Solution*

(i)

| $x$ | 1 | 2 | 3 | 4 |
|---|---|---|---|---|
| $x^2$ | 1 | 4 | 9 | 16 |
| $P(X = x)$ | 0.6 | 0.2 | 0.1 | 0.1 |

$E(X) = 1 \times 0.6 + 2 \times 0.2 + 3 \times 0.1 + 4 \times 0.1$
$= 1.7.$

$E(X^2) = 1 \times 0.6 + 4 \times 0.2 + 9 \times 0.1 + 16 \times 0.1$
$= 3.9.$

$Var(X) = E(X^2) - [E(X)]^2$
$= 3.9 - 1.7^2$
$= 1.01.$

(ii) $Var(7) = E(7^2) - [E(7)]^2$          *General result*
$= E(49) - [7]^2$          $Var(c) = 0$ for a constant $c$.
$= 49 - 49$          This result is obvious; a constant is
$= 0.$          constant and so can have no spread.

(iii) $Var(3X) = E[(3X)^2] - \mu^2$          *General result*
$= E(9X^2) - [E(3X)]^2$          $Var(aX) = a^2 Var(X)$.
$= 9E(X^2) - [3E(X)]^2$          Notice that it is $a^2$ and not $a$ on the right
$= 9 \times 3.9 - (3 \times 1.7)^2$          hand side, but that taking square roots
of both sides gives standard deviation
$= 35.1 - 26.01$          $(aX) = a \times$ standard deviation $(X)$ as
$= 9.09.$          you would expect from common sense.

(iv) $\text{Var}(3X + 7) = E[(3X + 7)^2] - [E(3X + 7)]^2$

$\qquad = E(9X^2 + 42X + 49) - [3E(X) + 7]^2$

$\qquad = E(9X^2) + E(42X) + E(49) - [3 \times 1.7 + 7]^2$

$\qquad = 9.E(X^2) + 42E(X) + 49 - 12.1^2$

$\qquad = 9 \times 3.9 + 42 \times 1.7 + 49 - 146.41$

$\qquad = 9.09.$

*General result*

$\text{Var}(aX + c) = a^2\text{Var}(X).$

Notice that the constant $c$ does not appear on the right hand side.

## Exercise 1C

**1**. The probability distribution of random variable $X$ is as follows:

| $x$ | 1 | 2 | 3 | 4 | 5 |
|---|---|---|---|---|---|
| $P(X = x)$ | 0.1 | 0.2 | 0.3 | 0.3 | 0.1 |

Find (i) $E(X)$;  (ii) $\text{Var}(X)$;  (iii) Verify that $\text{Var}(2X) = 4\text{Var}(X)$.

**2**. The probability distribution of a random variable $X$ is as follows:

| $x$ | 0 | 1 | 2 |
|---|---|---|---|
| $P(X = x)$ | 0.5 | 0.3 | 0.2 |

Find (i) $E(X)$; (ii) $\text{Var}(X)$; (iii) Verify that $\text{Var}(5X + 2) = 25\text{Var}(X)$.

**3**. Prove that $\text{Var}(aX - b) = a^2\,\text{Var}(X)$  where $a$ and $b$ are constants.

**4**. A coin is biased so that the probability of obtaining a tail is 0.75. The coin is tossed four times and the random variable $X$ is the number of tails obtained. Find (i) $E(2X)$;  (ii) $\text{Var}(3X)$.

**5**. Birds of a particular species lay either 0, 1, 2 or 3 eggs in their nests, with probabilities as shown in the following table.

| Number of eggs | 0 | 1 | 2 | 3 |
|---|---|---|---|---|
| Probability | 0.25 | 0.35 | 0.30 | $k$ |

Find

(i)  the value of $k$,
(ii) the expected number of eggs laid in a nest,
(iii) the standard deviation of the number of eggs laid in a nest.        [C]

**6**. A committee of 3 is to be selected at random from 3 men and 4 women. The number of men on the committee is the random variable $Y$.

Find (i) $E(Y)$; (ii) $\text{Var}(Y)$.

**7**. A discrete random variable $W$ has the distribution:

| $w$ | 1 | 2 | 3 | 4 | 5 | 6 |
|---|---|---|---|---|---|---|
| $P(W = w)$ | 0.1 | 0.2 | 0.1 | 0.2 | 0.1 | 0.3 |

Exercise 1c continued

Find the mean and variance of (i) $W + 7$;   (ii) $6W - 5$.

**8.** A cubical die is biased in such a way that the probability of scoring $n$ ($n = 1$ to 6) is proportional to $n$. Determine the mean value and variance of the score obtained in a single throw. What would be the mean and variance if the score showing were doubled?                [O&C]

**9.** The random variable $X$ is the number of heads obtained when 4 unbiased coins are tossed. Construct the probability distribution for $X$ and find:

(i) $E(X)$;   (ii) $Var(X)$;   (iii) $Var(3X + 4)$.

**10.** A shop sells red and blue refills for pens, and keeps them all in the same box. The box contains 5 red refills at £2.00 each and 4 blue ones at £1.60 each. A customer takes three refills out at random, not realising there are different colours in the box. Find the expectation and variance of the amount of money the customer spends.

**11.** A discrete random variable $X$ takes values 0, 1, 2, 4, 8 with probabilities as shown in the table.

| $x$ | 0 | 1 | 2 | 4 | 8 |
|---|---|---|---|---|---|
| $P(X = x)$ | $p$ | $\frac{1}{2}$ | $\frac{1}{4}$ | $\frac{1}{8}$ | $\frac{1}{16}$ |

(i)  Evaluate $p$.
(ii) Find $E(X)$ and $Var(X)$.
(iii) Write down the values of $E(2X + 3)$ and $Var(2X + 3)$.
(iv) $X_1$ and $X_2$ are two independent observations of $X$. Find $E(X_1 + X_2)$ and, using the value of $p$ found in (i) above, find $P(X_1 + X_2 = 2)$.     [C]

**12.** A box contains nine numbered balls. Three balls are numbered 3, four balls are numbered 4 and two balls are numbered 5.

Each trial of an experiment consists of drawing two balls without replacement and recording the sum of the numbers on them, which is denoted by $X$. Show that the probability that $X = 10$ is $1/36$, and find the probabilities of all other possible values of $X$.

Use your results to show that the mean of $X$ is $70/9$, and find the standard deviation of $X$.

Two trials are made. (The two balls in the first trial are replaced in the box before the second trial.) Find the probability that the second value of $X$ is greater than or equal to the first value of $X$.     [MEI]

**13.** A curiously shaped six-faced die produces scores, $X$, for which the probability distribution is given in the following table.

| $r$ | 1 | 2 | 3 | 4 | 5 | 6 |
|---|---|---|---|---|---|---|
| $P(X = r)$ | $k$ | $k/2$ | $k/3$ | $k/4$ | $k/5$ | $k/6$ |

Show that the constant $k$ is $^{20}/_{49}$, and find the mean and variance of $X$.

Show that, when this die is thrown twice, the probability of obtaining two equal scores is very nearly $^1/_4$. [MEI]

**14.** A bag contains four balls, numbered $2, 4, 6, 8$ but identical in all other respects. One ball is chosen at random and the number on it is denoted by $N$, so that $P(N = 2) = P(N = 4) = P(N = 6) = P(N = 8) = \frac{1}{4}$. Show that

$$\mu = E(N) = 5 \text{ and } \sigma^2 = \text{Var}(N) = 5.$$

Two balls are chosen at random one after the other, with the first ball being replaced after it has been drawn. Let $\overline{N}$ be the arithmetic mean of the numbers on the two balls. List the possible values of $\overline{N}$ and their probabilities of being obtained. Hence evaluate $E(\overline{N})$ and $\text{Var}(\overline{N})$. [MEI]

**15.** The discrete random variable $X$ takes values $-1, 0, 1$ with probabilities $\frac{1}{4}, \frac{1}{2}, \frac{1}{4}$ respectively. The variable $\overline{X}$ is the mean of a random sample of 3 values of $X$. Tabulate the probability distribution of $\overline{X}$, and use your values to calculate $\text{Var}(\overline{X})$. Hence verify that $\text{Var}(\overline{X}) = \frac{1}{3} \text{Var}(X)$ in this case. [C]

**16.** A random number generator in a computer game produces values which can be modelled by the discrete random variable $X$ with probability distribution given by

$$P(X = r) = kr! \quad r = 0, 1, 2, 3, 4,$$

where $k$ is a constant.

(i) Show that $k = ^1/_{34}$ and illustrate the probability distribution with a sketch.
(ii) Find the expectation and variance of $X$.

Two independent values of $X$ are generated. Let these values be $X_1$ and $X_2$.
(iii) Show that $P(X_1 = X_2)$ is a little greater than 0.5.
(iv) Given that $X_1 = X_2$, find the probability that $X_1$ and $X_2$ are each equal to 4. [MEI]

**17.** A traffic surveyor is investigating the lengths of queues at a particular set of traffic lights during the day time, but outside rush hours. He counts the number of cars, $X$, stopped and waiting when the lights turn green on 90 different occasions, with the following results:

| No. of cars, $x_i$ | 0 | 1 | 2 | 3 | 4 | 5 | 6 | 7 | 8 | 9+ |
|---|---|---|---|---|---|---|---|---|---|---|
| Frequency, $f_i$ | 3 | 9 | 12 | 15 | 15 | 15 | 11 | 8 | 2 | 0 |

(i)  Use these figures to estimate the probability distribution of the number of cars waiting when the lights turn green.
(ii)  Use your probability distribution to estimate the expectation and variance of $X$.

A colleague of the surveyor suggests that the probability distribution might be modelled by the expression

$$P(X=x_i) = kx_i(8-x_i)$$

(iii) Find the value of $k$.
(iv) Find the values of the expectation and variance of $X$ given by this model.
(v)  Do you think it is a good model?

**18.** A wine bar sells wine by the glass, and no other drinks during lunch time. It is usually very busy, sometimes even frantic. In order to help his planning, the owner commissions a mathematics student to do some research into the drinking habits of his customers. After observing a large number of people over a representative period, the student tells the owner '*The probability distribution of $X$, the number of drinks a customer will have is given by*

$$P(X=x_i) = \frac{(7-x_i)}{21} \quad \text{for } x_i = 1,2,\ldots,6$$
$$\text{otherwise } 0.'$$

The owner tells the student that he wants the information in a form he can understand.

(i)  Write the information in a form that the owner can indeed understand.
(ii)  Show that the probability distribution satisfies the conditions for $X$ to be a discrete random variable.
(iii) Calculate the expectation and variance of $X$.
(iv) Explain the significance to the bar's owner of whether the expectation and variance have larger small values.

The owner considers a possible new sales policy of selling only double glasses of wine (at double the price). He considers two possible models of how this might affect his custom:

(A) The same number of people come to the bar as before.

Those who previously would have had 2, 4 or 6 glasses, will now have 1, 2 or 3 double glasses, respectively.

Those who would have had 1, 3 or 5 glasses, round up their drinking by having 1, 2 or 3 double glasses respectively.

(B) Those who previously would have had 2, 4 or 6 glasses, will now have 1, 2 or 3 double glasses, respectively.

Half of those who would have had 1, 3 or 5 glasses, round up their drinking by having 1, 2 or 3 double glasses respectively; the other half round down by having 0, 1 or 2 double glasses.

Those who have 0 glasses no longer come to the bar at all.

(v) Calculate the mean and variance for both models.

(vi) Explain how you would interpret the answers to part (v) in terms of running the bar, and state what would you do if you were the owner.

# Investigations

Some of the work in the previous pages may have seemed quite theoretical to you. How does it apply to modelling real life situations?

Here are two practical investigations to help you to answer this for yourself. Make sure that you try at least one of them. It is important that you not only know the theory but have experience of how it works out in practice as well.

## Charity Stall

Invent a game for a charity stall where members of the public pay 10 pence a turn to select a card at random from a normal pack of 52. Decide for which outcomes the player is to win and what the winnings should be. Arrange for your game to make an expected profit of 3 pence per turn for the organiser.

As you play the game more and more times, keep a record of the results. You would expect the organiser's average profit per turn to get progressively closer to 3 pence. Does this really happen?

## Rolling a die

Work out the mean, $E(X)$, and variance, $Var(X)$, of the score, $X$, when a single die is rolled.

Roll a die 6 times and work out the values of the mean and variance of the scores you have obtained. How do they compare with the theoretical values you calculated?

Repeat the experiment but this time roll the die 60 times. Are your answers closer now?

Organise your friends so that, between you, the die is rolled 600 times. Are your answers even closer?

# 1

# Some Standard Discrete Probability Distributions

There are several standard discrete probability distributions which you will often need when modelling statistical data.

## The Binomial Distribution

You will recall that a binomial distribution results from $n$ independent trials of an experiment which has exactly two possible outcomes, often referred to as *success* (with probability $p$) and *failure* (with probability $q = 1 - p$). The probabilities $p$ and $q$ remain the same from one trial to the next, as for example, in rolling a die or tossing a coin.

For the binomial distribution, B($n,p$):

$$\text{Mean} = np; \quad \text{Variance} = npq = np(1\text{-}p).$$

(The proofs of these results are given in the appendix.)

**EXAMPLE**

Find the mean and variance of discrete random variable $X \sim$ B(8,0.2). (Notice the use of the symbol $\sim$ to mean 'is distributed as'.)

*Solution*

$$E(X) = np = 8 \times 0.2 = 1.6$$
$$\text{Var}(X) = npq = 8 \times 0.2 \times 0.8 = 1.28.$$

## The uniform distribution

This distribution occurs when all the possible outcomes are equally likely, as for example when a single die is rolled. There are six possible outcomes, each with the same probability, $1/6$, as shown in Figure 1.6.

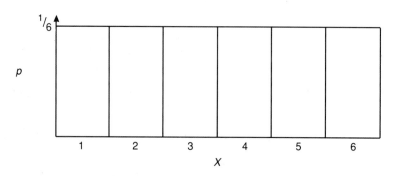

**Figure 1.6 Probability distribution of the score when a single die is rolled**

The mean and variance of this distribution are the same as those of the particular values the variable can take.

# The geometric distribution

If a sequence of independent trials is conducted, for each of which the probability of success is $p$ and that of failure $q$ ($q = 1 - p$), and the random variable $X$ is the number of the trial on which the first success occurs, then $X$ has a *geometric distribution*.

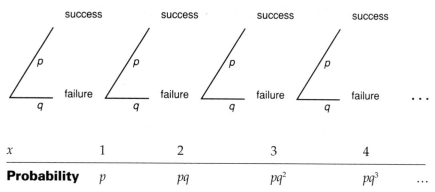

| $x$ | 1 | 2 | 3 | 4 | |
|---|---|---|---|---|---|
| **Probability** | $p$ | $pq$ | $pq^2$ | $pq^3$ | ... |

Thus the distribution is given by

$$P(X = x) = pq^{x-1} \quad \text{for } x \geq 1$$

An example of this is the probability distribution of the number of times you have to roll a die until the number 6 shows.

For this distribution: Mean = $1/p$; Variance = $q/(p^2)$

## Exercise 1D

1. The random variable, $X$, is the number of heads when a fair coin is tossed 10 times.

   (i) Copy and complete this table giving the probability distribution of $X$.

   | $x_i$ | 0 | 1 | 2 | ... | 10 |
   |---|---|---|---|---|---|
   | $P(X=x_i)$ | $^1/_{1024}$ | $^{10}/_{1024}$ | | | $^1/_{1024}$ |

   (ii) Use these figures to find $E(X)$ and $Var(X)$.
   (iii) Show that you get the same values by using $E(X)=np$ and $Var(X)=npq$.

2. The random variable, $Y$, is the number of sixes when 5 fair dice are rolled.

   (i) Copy and complete this table giving the probability distribution of $Y$.

   | $Y_i$ | 0 | 1 | 2 | ... | 5 |
   |---|---|---|---|---|---|
   | $P(Y=y_i)$ | 0.4019 | | | | |

   (ii) Use these figures to find $E(Y)$ and $Var(Y)$.
   (iii) Show that you get the same values by using $E(Y)=np$ and $Var(Y)=npq$.

**3.** An applicant for a post in a casino is asked to demonstrate that she can toss a coin fairly. She is to carry out 20 sets of trials, each consisting of 10 tosses of the coin. In each the number of heads is recorded.

(i) What are the expectation and variance of the numbers of heads in a set of 10 tosses?

In her 20 sets of trials, the applicant gets the following numbers of heads:

3  5  4  8  7    7  5  5  5  4    3  4  4  6  6    7  5  6  4  5

(ii) Calculate the mean and variance of these data.

(iii) Do you think the applicant is tossing the coin fairly?

**4.** At a charity fund-raising dinner, each guest is to be asked to roll a fair die and give £5 for each spot on the uppermost face of the die. The random variable $Y$ is the amount raised per person in this way.

(i) Write out the probability distribution of $Y$.

(ii) Find the expectation and the standard deviation of $Y$.

There are 100 guests at the dinner. When the money has been collected from the die rolling it is found to amount to £2750.

(iii) What do you think has happened?

**5.** A couple decide they will have children until they have a daughter, and then stop. Assuming that each child is equally likely to be a girl or a boy,

(i) find the expectation of the number of children they will have.

Their friends decide to have children until they have one of each sex.

(ii) What is the expectation of their number of children?

(iii) Find, for each couple, the probability that they will end up with at least four children.

**6.** A box contains 3 red balls and 2 green balls. A ball is drawn out at random and then replaced until a green ball is obtained. If $X$ represents the number of draws required to pick a green ball,

(i) describe the probability distribution of $X$;

(ii) find $E(X)$ and $Var(X)$.

**7.** In a promotion campaign a breakfast cereal company includes between 1 and 4 vouchers in each of its packets, the number, $X$, being random with all four possibilities equally likely.

(i) Write down the probability distribution of $X$ and find $E(X)$ and $Var(X)$.

A family buy two packets of the cereal and receive $Y$ vouchers.

(ii) Write down the probability distribution of $Y$ and find $E(Y)$ and $Var(Y)$.

The promotion lasts for only a few months and anyone who gets 6 or more vouchers is entitled to exchange them for a 'free' gift.

*Exercise 1d continued*

> (iii) What is the least number of packets that someone must buy in order to have a probability of $\frac{1}{2}$ or more of qualifying for the gift?
> (iv) Mr. MacTaggart buys three packets. What is the expectation of the number of gifts he will receive?

8. A coin is biased in such a way that, on any throw, P(head) $= p$ and P(tail) $= q$, where $p + q = 1$. The random variable $X$ denotes the number of heads resulting from three throws of this coin. Tabulate the probability distribution of $X$, and show from first principles (i.e. without quoting any results relating to the binomial distribution) that $E(X) = 3p$ and that $Var(X) = 3pq$.

In an experiment, three throws of the coin were repeated 1000 times. The numbers of times each value of $X$ occurred are shown in the table.

| Number of heads | 0 | 1 | 2 | 3 |
|---|---|---|---|---|
| Frequency | 90 | 329 | 412 | 169 |

Calculate the mean number of heads from these figures, and use it to estimate the value of $p$.

Hence estimate the probability that five throws of this coin will result in at least one head. [C]

# Summary

For a discrete random variable $X$ which can assume only the values

$$x_1, x_2, ..., x_n$$

with probabilities $p_1, p_2, ..., p_n$ respectively,

$$\Sigma p_i = 1 \quad p_i \geq 0$$
$$E(X) = \Sigma x_i p_i = \Sigma x_i P(X = x_i)$$
$$E(g[X]) = \Sigma g[x_i] p_i = \Sigma g[x_i] P(X = x_i)$$
$$Var(X) = E(X^2) - [E(X)]^2.$$

For any discrete random variable $X$, and constants $a$ and $c$:

$$E(c) = c$$
$$E(aX) = aE(X)$$
$$E(aX + c) = aE(X) + c$$
$$E(f[X] + g[X]) = E(f[X]) + E(g[X])$$
$$Var(c) = 0$$
$$Var(aX) = a^2 Var(X)$$
$$Var(aX + c) = a^2 Var(X).$$

# The Poisson Distribution

*If something can go wrong, sooner or later it will go wrong.*

Murphy's Law

THE AVONFORD STAR

# Rare Disease Blights Town

**Chemical Plant Blamed**

A rare disease is attacking residents of Avonford. In the last year alone 5 people have been diagnosed as suffering from it. This is over three times the national average.

The disease (known as *Palfrey's Condition*) causes nausea and fatigue. One sufferer James Louth (32), of 52 Harpers Lane, has been unable to work for the past six months. His wife Muriel (29) said "I am worried sick, James has lost his job and I am frightened that the children [Mark (4) and Samantha (2)] will catch it."

Mrs Louth blames the chemical complex on the Industrial Estate for the disease. "There were never any cases before Avonford Chemicals arrived."

Local environmental campaigner Roy James supports Mrs. Louth. "I warned the Local Council when planning permission was sought that this would mean an increase in this sort of illness. Normally we would expect 1 case in every 40,000 of the population in a year."

*Muriel Louth believes the local chemical plant could destroy her family's lives*

Avonford Chemicals spokesperson, Julia Millward said, "We categorically deny that our plant is responsible for the disease. Our record on safety is very good. None of our staff has had the disease. In any case five cases in a population of 60,000 can hardly be called significant."

The expected number of cases is 60 000 × 1/40 000 or 1.5, so 5 does seem rather high.

Do you think that the chemical company is to blame or do you think people are just looking for an excuse to attack it? How do you decide between the two points of view? Is 5 really that large a number of cases anyway?

The situation could be modelled by the binomial distribution. The probability of someone getting the virus in any year is $1/40\,000$ and so that of not getting it is $1-1/40\,000 = 39\,999/40\,000$.

So the probability of 5 cases among 60 000 people (and so 59 995 people not getting it) is given by

$$^{60\,000}C_5(39\,999/40\,000)^{59\,995}(1/40\,000)^5, \text{ about } 0.0141.$$

What you really want to know however is not the probability of exactly 5 cases but that of 5 or more cases. If that is very small, then perhaps something unusual did happen in Avonford last year.

You can find by finding the probability of up to and including 4 cases, and subtracting it from 1.

The probability of up to and including 4 cases is given by this forbidding expression:

$$(39\,999/40\,000)^{60\,000} \qquad\qquad \text{0 cases}$$
$$+\ ^{60\,000}C_1(39\,999/40\,000)^{59\,999}(1/40\,000) \qquad \text{1 case}$$
$$+\ ^{60\,000}C_2(39\,999/40\,000)^{59\,998}(1/40\,000)^2 \qquad \text{2 cases}$$
$$+\ ^{60\,000}C_3(39\,999/40\,000)^{59\,997}(1/40\,000)^3 \qquad \text{3 cases}$$
$$+\ ^{60\,000}C_4(39\,999/40\,000)^{59\,996}(1/40\,000)^4 \qquad \text{4 cases.}$$

It is messy but you can evaluate it on your calculator. It comes out to be

$$0.223 + 0.335 + 0.251 + 0.126 + 0.047 = 0.981$$

(The figures are written to 3 decimal places but more places were used for the addition.)

So the probability of 5 or more cases in a year is $1 - 0.981 = 0.019$.

It is unlikely, with a probability of about $1/50$, but certainly could happen, see figure 2.1.

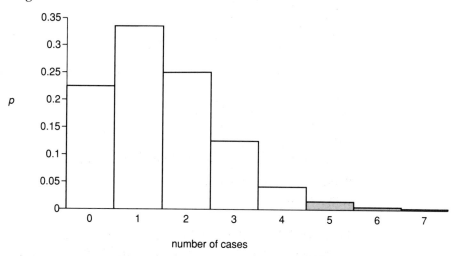

Figure 2.1 Probability distribution B (60 000, 1/40 000)

Two other points are worth making. First, the binomial model assumes the trials are independent. If this disease is at all infectious, that certainly would not be the case. Second, there is no evidence at all to link this disease with the Avonford Chemicals. There are many other possible explanations.

# Approximating the binomial terms

Although it was possible to do the calculation using results derived from the binomial distribution, it was distinctly cumbersome. In this section you will see how the calculations can be simplified, a process which turns out to be unexpectedly profitable. The work that follows depends upon the facts that the event is rare but there are many opportunities for it to occur: $p$ is small and $n$ is large.

Start by looking at the last term, the probability of 4 cases of the disease. This is

$$^{60\,000}C_4(39\,999/40\,000)^{59\,996}(1/40\,000)^4$$

Writing

$$^{60\,000}C_4 \text{ as } \frac{60\,000 \times 59\,999 \times 59\,998 \times 59\,997 \times 59\,996}{1 \times 2 \times 3 \times 4}$$

makes the expression for the probability

$$\frac{60\,000 \times 59\,999 \times 59\,998 \times 59\,997 \times 59\,996 \times (39\,999)^{59\,996}}{1 \times 2 \times 3 \times 4 \times (40\,000)^{59\,996} \times (40\,000)^4}$$

This can be simplified a great deal by making the following approximations:

Write 59 999, 59 998, 59 997 and 59 996 as 60 000;

write 39 999 as 40 000

(except where they appear as powers).

The expression now becomes

$$\frac{60\,000 \times 60\,000 \times 60\,000 \times 60\,000 \times 60\,000 \times (40\,000)^{59\,996}}{1 \times 2 \times 3 \times 4 \times (40\,000)^{59\,996} \times (40\,000)^4}$$

which cancels down to

$$\frac{1}{4!} \times \left(\frac{60\,000}{40\,000}\right)^4$$

or $$\frac{(1.5)^4}{4!}$$

Notice that the number 1.5 has arisen as 60 000/40 000 so it is just the expectation, or mean, of the number of cases.

Carrying out the same procedure on the other terms gives the series

$$1 + 1.5 + \frac{(1.5)^2}{2!} + \frac{(1.5)^3}{3!} + \frac{(1.5)^4}{4!}$$

Cases       0    1    2     3     4

This result is very neat but it is immediately obvious that the approximations have caused a problem. For this to be a probability distribution it must sum to 1 but here the first term alone is 1 and the total sum is now clearly greater than 1. Each term needs to be divided by the total sum of this series (not just the first five terms given above).

Fortunately this is in fact a well known series, the exponential series, $e^x$.

$$e^x = 1 + x + x^2/2! + x^3/3! + x^4/4! + \ldots$$

So the probability distribution for the number of cases of people with the virus is found by dividing the terms by $e^{1.5}$, or multiplying them by $e^{-1.5}$, which is the same thing. This gives this distribution:

| Number of cases | 0 | 1 | 2 | 3 | 4 | ... |
|---|---|---|---|---|---|---|
| Probability | $e^{-1.5}$ | $e^{-1.5}1.5$ | $e^{-1.5}\dfrac{(1.5)^2}{2!}$ | $e^{-1.5}\dfrac{(1.5)^3}{3!}$ | $e^{-1.5}\dfrac{(1.5)^4}{4!}$ | ... |

and in general for $r$ cases the probability is $\quad e^{-1.5}\dfrac{(1.5)^r}{r!}$

## Accuracy

These expressions are clearly much simpler than those involving binomial coefficients. How accurate are they? The following table compares the results from the two methods, given to 6 decimal places.

| No. of cases | Probability | |
|---|---|---|
| | Exact binomial method | Approximate method |
| 0 | 0.223126 | 0.223130 |
| 1 | 0.334697 | 0.334695 |
| 2 | 0.251025 | 0.251021 |
| 3 | 0.125512 | 0.125511 |
| 4 | 0.047066 | 0.047067 |

You will see that the agreement is very good; there are no differences until the sixth decimal places.

## The Poisson distribution

What started out as a search for an easy way to calculate terms in a binomial distribution has ended up with terms which are so different that they are seen as a completely different distribution, called the *Poisson distribution*.

In the example the distribution had a mean of 1.5, and the general term was

given by:
$$P(X=r) = e^{-1.5}\frac{(1.5)^r}{r!}$$

where the discrete random variable $X$ denoted the number of cases of the disease.

This can be generalised to the Poisson distribution with mean $\lambda$ (pronounced 'lambda') for which
$$P(X=r) = e^{-\lambda}\frac{\lambda^r}{r!}$$

**NOTES**

1. *The mean, $\lambda$, is usually called the population parameter.*

2. *e is the base of natural logarithms, is 2.718 28...; like $\pi$, it does not terminate. $e^x$ is a function on your calculator.*

3. *The upper case letter X is a random variable, and the lower case r is a particular value that X can take.*

4. *The shape of the Poisson distribution depends on the value of the parameter, $\lambda$. If $\lambda$ is small the distribution has positive skew, but as $\lambda$ increases the distribution becomes progressively more symmetrical, see figure 2.2.*

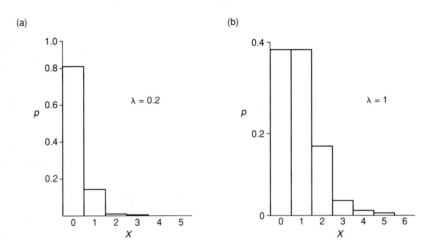

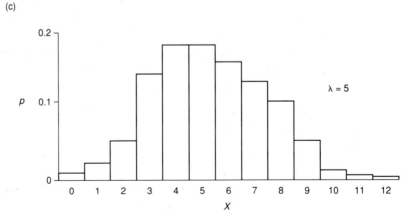

**Figure 2.2 The shape of the Poisson distribution for (a) $\lambda = 0.2$ (b) $\lambda = 1$ (c) $\lambda = 5$**

In the previous few pages, you have been introduced to the Poisson distribution as a very good approximation to the binomial distribution. However it is much more significant than that. It is a distribution in its own right which can be used to model situations which are clearly not binomial, where you cannot state the number of trials or the probabilities of success and failure.

## Conditions under which the Poisson distribution may be used

The Poisson distribution is generally thought of as the probability distribution for the number of occurrences of a *rare event*.

The conditions under which it may be used are as follows:

### 1. *As an approximation to the binomial distribution*

The Poisson distribution may be used as an approximation to the binomial distribution, $B(n,p)$, when

    (i)    $n$ is large
    (ii)    $p$ is small (and so the event is rare)
    (iii)    $np$ is not large

In addition you would not usually want to use the Poisson distribution for large values of $np$, so in practice (but not in theory) a third condition may be applied:

In the case of the virus, $n$ was large at 60 000, $p$ small at $1/40\ 000$. In addition $np$, 1.5, was not large.

It is also necessary that the trials are random and independent; otherwise the distribution to be approximated would not be binomial in the first place.

### 2. *As a distribution in its own right*

There are many situations in which the mean number of occurrences is known (or can easily be found) but it is not possible, or even meaningful, to give values to the number of trials, $n$, or the probability of success, $p$. You can for example find the mean number of goals per team in league football matches, or the mean number of telephone calls received per minute at an exchange, but you cannot say how many goals are not scored or telephone calls not received. The concept of a trial, with success or failure as possible outcomes, is not appropriate to these situations.

They may, however, be modelled by the Poisson distribution, provided that the occurrences are

    (i)    random;
    (ii)    independent;
    (iii)    that events occur with uniform likelihood over the interval, as in the example that follows.

It would be unusual to use the Poisson distribution with a population mean greater than about 20, because easier methods exist.

**EXAMPLE** The number of defects in a wire cable can be modelled by the Poisson distribution with a mean of 4 defects per km.

What is the probability that a single km. will have exactly 2 defects?

*Solution*

Let $X$ = the number of defects per km.,      $P(X=2) = e^{-4}\dfrac{4^2}{2!}$

$$= 0.147.$$

## Notation

The notation '$X \sim \text{Poisson}(6)$' means that the random variable $X$ has the Poisson distribution with parameter (or mean) 6.

So the probability that $X$ takes the value 8 is given by $e^{-6}6^8/8!$ and the probability that it takes a general value $r$ by $e^{-6}6^r/r!$

## Calculating Poisson Distribution Probabilities

In the example about the disease afflicting Avonford, you had to work out $P(X\geq5)$. To do this you used

$$P(X\geq5) = 1 - P(X\leq4)$$

which of course saved you having to work out all the probabilities for 5 or more occurrences and adding them together. Such calculations can take a long time even though the terms eventually get smaller and smaller, so that after some time you will have gone far enough for the accuracy you require and may stop.

However suppose you had $X \sim \text{Poisson}(8)$ and wanted to find $P(X\leq7)$.

$$P(X\leq7) = P(X=0) + P(X=1) + P(X=2) + P(X=3) + \dots + P(X=7)$$

This involves summing eight probabilities and so is itself rather tedious. Here are two ways of cutting down on the amount of work, and so on the time you take.

### 1. *Recurrence relations*

Recurrence relations allow you to use the term you have got to work out the next one. For the Poisson distribution with parameter $\lambda$,

$$P(X=0) = e^{-\lambda}$$      You must use your calculator to find this first term.

$$P(X=1) = e^{-\lambda}\lambda \quad = \lambda P(X=0)$$      Multiply the previous term by $\lambda$.

$$P(X=2) = e^{-\lambda}\dfrac{\lambda^2}{2!} = \dfrac{\lambda}{2}P(X=1)$$  Multiply the previous term by $\lambda/2$.

$$P(X=3) = e^{-\lambda} \frac{\lambda^3}{3!} = \frac{\lambda}{3} P(X = 2) \quad \text{Multiply the previous term by } \lambda/3.$$

$$P(X=4) = e^{-\lambda} \frac{\lambda^4}{4!} = \frac{\lambda}{4} P(X = 3) \quad \text{Multiply the previous term by } \lambda/4.$$

In general, you can find $P(X=r)$ by multiplying your previous probability, $P(X=r\text{-}1)$, by $\lambda/r$. You would expect to hold the latest value on your calculator and keep a running total in memory.

Setting this out on paper with $\lambda = 1.5$ (the figure from the virus example) gives these figures:

| No. of cases, $r$ | Conversion | $P(X=r)$ | Running total, $P(X\leq r)$ |
|---|---|---|---|
| 0 | | 0.223130 | 0.223130 |
| | $\times$ 1.5 | | |
| 1 | | 0.334695 | 0.557825 |
| | $\times$ 1.5/2 | | |
| 2 | | 0.251021 | 0.808846 |
| | $\times$ 1.5/3 | | |
| 3 | | 0.125511 | 0.934357 |
| | $\times$ 1.5/4 | | |
| 4 | | 0.047067 | 0.981424 |

## 2. Cumulative Poisson Probability Tables

You can also find probabilities like $P(X\leq 4)$ by using *Cumulative Poisson Probability Tables*. You can see how to do this by looking at the extract from the tables below. For $\lambda=1.5$ and $x=4$ this gives you the answer 0.9814.

| $x$ \ $\lambda$ | 1.00 | 1.10 | 1.20 | 1.30 | 1.40 | 1.50 | 1.60 | 1.70 | 1.80 | 1.90 |
|---|---|---|---|---|---|---|---|---|---|---|
| 0 | 0.3679 | 0.3329 | 0.3012 | 0.2725 | 0.2466 | 0.2231 | 0.2019 | 0.1827 | 0.1653 | 0.1496 |
| 1 | 0.7358 | 0.6990 | 0.6626 | 0.6268 | 0.5918 | 0.5578 | 0.5249 | 0.4932 | 0.4628 | 0.4337 |
| 2 | 0.9197 | 0.9004 | 0.8795 | 0.8571 | 0.8335 | 0.8088 | 0.7834 | 0.7572 | 0.7306 | 0.7037 |
| 3 | 0.9810 | 0.9743 | 0.9662 | 0.9569 | 0.9463 | 0.9344 | 0.9212 | 0.9068 | 0.8913 | 0.8747 |
| 4 | 0.9963 | 0.9946 | 0.9923 | 0.9893 | 0.9857 | 0.9814 | 0.9763 | 0.9704 | 0.9636 | 0.9559 |
| 5 | 0.9994 | 0.9990 | 0.9985 | 0.9978 | 0.9968 | 0.9955 | 0.9940 | 0.9920 | 0.9896 | 0.9868 |
| 6 | 0.9999 | 0.9999 | 0.9997 | 0.9996 | 0.9994 | 0.9991 | 0.9987 | 0.9981 | 0.9974 | 0.9966 |
| 7 | 1.0000 | 1.0000 | 1.0000 | 0.9999 | 0.9999 | 0.9998 | 0.9997 | 0.9996 | 0.9994 | 0.9992 |
| 8 | ..... | ..... | ..... | 1.0000 | 1.0000 | 1.0000 | 1.0000 | 0.9999 | 0.9999 | 0.9998 |
| 9 | ..... | ..... | ..... | ..... | ..... | ..... | ..... | 1.0000 | 1.0000 | 1.0000 |

*Extract from the cumulative Poisson probability tables x=4, $\lambda=1.5$ gives P(x$\leq$4) = 0.9814*

If you wanted to find $P(X\geq 5)$, you would use

$$P(X\geq 5) = 1 - P(X\leq 4)$$

$$= 1 - 0.9814$$

$$= 0.0186.$$

# The variance of the Poisson distribution

If you think of the Poisson distribution as an approximation to the binomial, $B(n,p)$, when $p$ is small and $n$ is large, you can find an expression for the variance by approximating that for the binomial distribution.

For the binomial distribution, $B(n,p)$:

$$\text{Mean} = np \quad \text{Variance} = np(1-p) = npq.$$

When $p$ is small, $q = 1-p$ is approximately equal to 1.

Writing $q=1$ in    Variance $= npq$

gives    Variance $= np$.

You will see that this is just the same as the mean, giving the somewhat surprising result that for a Poisson distribution

$$\text{Variance} = \text{mean}.$$

Both variance and mean are given the symbol $\lambda$ and called the *population parameter*.

So if you have some data where the mean and variance are very similar then you may possibly be able to use the Poisson distribution as a model. In real life you will not find the mean and variance to be exactly the same. Remember that you are using the Poisson distribution as a model and it is unlikely to be a perfect fit.

**NOTES**

1. *You should realise that this is not a rigid proof of the expression for the variance of the Poisson distribution. For that you would need to start, not with the binomial distribution, but with the Poisson probability:*

$$P(X=r) \;=\; e^{-\lambda}\,\frac{\lambda^r}{r!}$$

*This proof is somewhat longer and is given in the Appendix on page 127.*

2. *You may have wondered earlier why the term* parameter *was used rather than* mean, *and if you were perceptive, why it was given the symbol $\lambda$ rather than $\mu$. Since the variance and mean are both the same, either of them can be used to tell you all you want to know about a Poisson distribution and so either is a parameter. To say that either one of the mean, $\mu$, or variance, $\sigma^2$, is the parameter would be to give one of them greater status than the other. So a new symbol, $\lambda$, is used.*

**EXAMPLE** It is known that nationally one person in a thousand is allergic to a particular chemical used in making a wood preservative. A firm that makes this wood preservative employs 500 people in one of its factories.

2

(i) What is the probability that more than two people at the factory are allergic to the chemical?

(ii) What assumption are you making?

*Solution*

(i) Let $X$ be the number of people in a random sample of 500 who are allergic to the chemical.

$$X \sim B(500, 0.001) \qquad n=500 \quad p=0.001$$

Since $n$ is large and $p$ is small, the Poisson approximation to the binomial is appropriate.

$$\begin{aligned} \lambda &= np \\ &= 500 \times 0.001 \\ &= 0.5 \end{aligned}$$

Consequently
$$\begin{aligned} P(X=r) &= e^{-\lambda} \frac{\lambda^r}{r!} \\ &= e^{-0.5} \frac{0.5^r}{r!} \end{aligned}$$

$$\begin{aligned} P(X>2) &= 1 - P(X \le 2) \\ &= 1 - [P(X=0) + P(X=1) + P(X=2)] \\ &= 1 - \left[ e^{-0.5} + e^{-0.5} 0.5 + e^{-0.5} \frac{0.5^2}{2} \right] \\ &= 1 - [0.6065 + 0.3033 + 0.0758] \\ &= 1 - 0.9856 \\ &= 0.014 \end{aligned}$$

This figure could have been found in Cumulative Poisson Probability tables.

(ii) The assumption made is that people with the allergy are just as likely to work in the factory as those without the allergy. In practice this seems rather unlikely: you would not stay in a job that made you unhealthy.

**EXAMPLE** Jasmit is considering buying a telephone answering machine. He has one for 5 days' free trial and finds that 22 messages are left on it. Assuming that this is typical of the use it will get if he buys it, find:

(i) the mean number of messages per day;

(ii) the probability that on one particular day there will be exactly 6 messages;

(iii) the probability that on one particular day there will be more than 8 messages.

*Solution*

(i) Converting the total for 5 days to the mean for a single day gives

$$\text{Daily mean} = \frac{22}{5} = 4.4 \text{ messages per day}$$

(ii) Calling $X$ the number of messages per day,

$$P(X=6) = e^{-4.4} \frac{4.4^6}{6!}$$

$$= 0.124$$

(iii) Using the Cumulative Poisson Probability tables gives

$$P(X \leq 6) = 0.8436$$

and so
$$P(X>6) = 1 - 0.8436$$
$$= 0.1564.$$

*Exercise 2A*

**1.** If $X \sim \text{Poisson}(2)$, calculate:    (i) $P(X=1)$;    (ii) $P(X=4)$.

**2.** If $Y \sim \text{Poisson}(5)$, calculate:    (i) $P(Y=5)$;    (ii) $P(Y=7)$.

**3.** If $Z \sim \text{Poisson}(2.5)$, calculate:    (i) $P(Z=0)$;    (ii) $P(Z=3)$;    (iii) $P(Z=5)$.

**4.** If $W \sim \text{Poisson}(3)$, calculate:    (i) $P(W=0)$;    (ii) $P(W=1)$;    (iii) $P(W=2)$;    (iv) $P(W \leq 2)$;    (v) $P(W>2)$.

**5.** If $X \sim \text{Poisson}(4.4)$, calculate:    (i) $P(X \leq 3)$;    (ii) $P(X>3)$.

**6.** If $Y \sim \text{Poisson}(8)$, calculate:    (i) $P(Y \leq 5)$;    (ii) $P(Y>5)$.

**7.** If $Z \sim \text{Poisson}(9.5)$, calculate:    (i) $P(Z \leq 6)$;    (ii) $P(Z \geq 7)$.

**8.** A typesetter makes 1500 mistakes in a book of 500 pages. On how many pages would you expect to find (i) 0; (ii) 1; (iii) 2; (iv) 3 or more mistakes? State and comment on any assumptions in your working.

**9.** During the 1978–79 season Liverpool F.C. scored the following numbers of goals in their league matches:

| Goals | 0 | 1 | 2 | 3 | 4 | 5 | 6 |
|---|---|---|---|---|---|---|---|
| Matches | 7 | 11 | 10 | 8 | 3 | 1 | 2 |

(i)  How many league matches did Liverpool play that season?
(ii)  Calculate the mean and variance of the number of goals per match.
(iii) Using the mean you calculated in part (i) as parameter in the Poisson distribution, calculate the probabilities of $0, 1, ..., 5, 6$ or more goals in any match.
(iv) In how many matches would you expect Liverpool to have scored $0, 1, 2, ...$ goals, according to the Poisson model?
(v)  State, with reasons, whether you consider this to be a good model.

**10.** In a country the mean number of deaths per year from lightning strike is 2.2. Find the probabilities of (i) 0; (ii) 1; (iii) 2; (iv) more than 2 deaths from lightning strike in any particular year.

In a neighbouring country, it is found that one year in twenty no-one dies from lightning strike.

(v)  What is the mean number of deaths per year in that country from lightning strike?

**11.** A machine that puts an anchovy stuffing into olives fails to get the stuffing into one in every 50 olives. Find the probability that in a tub of 100 stuffed olives:

(i)  all the olives are correctly stuffed;
(ii)  there is exactly 1 unstuffed olive.

The olives are sold in packs of 250 tubs.

(iii) How many tubs in such a pack would you expect to have 4 or more unstuffed olives?

One person in a thousand complains if there are 4 or more unstuffed olives in a tub. How many complaints a year would you expect from a shop that orders 6 packs a month?

**12.** A manufacturer of rifle ammunition tests a large consignment for accuracy by firing 500 batches, each of 20 rounds, from a fixed rifle at a target. Those rounds that fall outside a marked circle on the target are classified as *misses*. For each batch of 20 rounds the number of misses is counted.

| Misses, $X$ | 0 | 1 | 2 | 3 | 4 | 5 | 6–20 |
|---|---|---|---|---|---|---|---|
| Frequency | 230 | 189 | 65 | 15 | 0 | 1 | 0 |

(i)  Estimate the mean number of misses per batch.
(ii)  Use your mean to estimate the probability of a batch producing $0, 1, 2, 3, 4$ and 5 misses using the Poisson distribution as a model.
(iii) Use your answers to part (ii) to estimate expected frequencies of $0, 1, 2, 3, 4$ and 5 misses per batch in 500 batches, and compare your answers with those actually found.

*Exercise 2a continued*

    (iv) Do you think the Poisson distribution is a good model for this situation?

**13.** A survey in a town's primary schools has indicated that 5% of the pupils have severe difficulties with reading. If the primary school pupils were allocated to the secondary schools at random, estimate the probability that a secondary school with an intake of 200 pupils will receive:

    (i)  no more than 8 pupils with severe reading difficulties;

    (ii) more than 20 pupils with severe reading difficulties.

**14.** Fanfold paper for computer printers is made by putting perforations every 30 cm in a continuous roll of paper. A box of fanfold paper contains 2000 sheets. State the length of the continuous roll from which the box of paper is produced.

The manufacturers claim that faults occur at random and at an average rate of 1 per 240 metres of paper. State an appropriate distribution for the number of faults per box of paper. Find the probability that a box of paper has no faults and also the probability that it has more than 4 faults.

Two copies of a report which runs to 100 sheets per copy are printed on this sort of paper. Find the probability that there are no faults in either copy of the report and also the probability that just one copy is faulty.

[MEI]

**15.** A firm investigated the number of employees suffering injuries whilst at work. The results recorded below were obtained for a 52-week period:

| Number of employees injured in a week | 0 | 1 | 2 | 3 | 4 or more |
|---|---|---|---|---|---|
| Number of weeks | 31 | 17 | 3 | 1 | 0 |

Give reasons why one might expect this distribution to approximate to a Poisson distribution. Evaluate the mean and variance of the data and explain why this gives further evidence in favour of a Poisson distribution.

Using the calculated value of the mean, find the theoretical frequencies of a Poisson distribution for the number of weeks in which 0, 1, 2, 3, 4 or more, employees were injured.

**16.** In what circumstances may the Poisson distribution be used as an approximation to the Binomial distribution with $N$ trials and a probability of success of $p$ in each trial?

350 raisins are put into a mixture which is well stirred and made into 100 small buns. Estimate how many of these buns will:

    (i)  be without raisins,

    (ii) contain 5 or more raisins.

In a second batch of 100 buns, exactly one has no raisins in it. Estimate the total number of raisins in the second mixture.

**17.** A count was made of the number of red blood corpuscles in each of the 64 compartments of a haemocytometer with the following results:

| Number of corpuscles | 2 | 3 | 4 | 5 | 6 | 7 | 8 | 9 | 10 | 11 | 12 | 13 | 14 |
|---|---|---|---|---|---|---|---|---|---|---|---|---|---|
| Frequency | | 1 | 5 | 4 | 9 | 10 | 10 | 8 | 6 | 4 | 3 | 2 | 1 | 1 |

Estimate the mean and variance of the number of red blood corpuscles per compartment. Explain how the values you have obtained support the view that these data are a sample from a Poisson population.

Write down an expression for the theoretical frequency with which compartments containing 5 red blood corpuscles should be found, assuming this to be obtained from a Poisson population with mean 7. Evaluate this frequency to two decimal places. [MEI]

**18.** At a busy intersection of roads, accidents requiring the summoning of an ambulance occur with a frequency, on average, of 1.8 per week. These accidents occur randomly, so that it may be assumed that they follow a Poisson distribution.

(i) Calculate the probability that there will not be an accident in a given week.

(ii) Calculate the smallest integer $n$ such that the probability of more than $n$ accidents in a week is less than 0.02.

(iii) Calculate the probability that there will not be an accident in a given fortnight.

(iv) Calculate the largest integer $k$ such that the probability that there will not be an accident in $k$ successive weeks is greater than 0.0001. [AEB]

**19.** A ferry takes cars and small vans on a short journey from an island to the mainland. On a representative sample of weekday mornings, the numbers of vehicles, $X$, on the 8 am sailing were as follows:

| 20 | 24 | 24 | 22 | 23 | | 21 | 20 | 22 | 23 | 22 |
|---|---|---|---|---|---|---|---|---|---|---|
| 21 | 21 | 22 | 21 | 23 | | 22 | 20 | 22 | 21 | 23 |

(i) Show that $X$ does not have a Poisson distribution.

In fact 20 of the vehicles belong to commuters who use that sailing of the ferry every weekday morning. The random variable $Y$ is the number of vehicles other than those 20 who are using the ferry.

(ii) Investigate whether $Y$ may reasonably be modelled by a Poisson distribution.

The ferry can take 25 vehicles on any journey.

(iii) On what proportion of days would you expect at least one vehicle to be unable to travel on this particular sailing of the ferry because there was no room left, and so wait for the next one?

# Experiments

The Poisson distribution provides a remarkably accurate model for many situations. Here are some experiments for you to try, but before you start there are a number of points to keep in mind.

## 1.  Modelling

Remember that if your results do not give a Poisson distribution, then that tells you it is not a good model. It does not mean that your experiment has failed, merely that the model is inappropriate.

## 2.  Independence

Each occurrence must be independent. If, for example, you collect data on the numbers of cars passing a point just beyond some traffic lights say every 30 seconds, the data will not be independent.

## 3.  Statistics experiments

In Statistics the word *experiment* means a process which results in the collection of data; this is a slightly different, and sometimes less active, meaning from that elsewhere. You can for instance do a Statistics experiment sitting at your desk with a newspaper in front of you.

Collect data on one of the following:

(i)   The number of goals scored by teams in the Football League on a single Saturday.

(ii)  The number of telephone calls coming into a switchboard in a 1 minute period. (You may wish to vary the time period to, say, 5 minutes if the calls are infrequent).

(iii) The number of cars passing a suitable point in a 1 minute period.

(iv) Data relating to a situation of your own choice which you think could be modelled well by the Poisson distribution.

When you have collected the data, go through the following steps.

1.   Work out the mean and the variance and check that they are approximately equal.

2.   Use the mean to work out the Poisson probability distribution and a suitable set of expected frequencies.

3.   Compare these expected frequencies with your observations.

**N O T E**     *In the first three cases, you know the number of events that have occurred, but not how many have not occurred, so you cannot possible use the binomial distribution. It makes no sense to ask 'How many goals were not scored?' You can estimate the mean, np, but not n or p on their own.*

# The sum of two or more Poisson distributions

It is often the case that you wish to add two or more Poisson distributions together.

**EXAMPLE**

A rare disease causes the deaths, on average, of 3.8 people per year in England, 0.8 in Scotland and 0.5 in Wales. As far as is known the disease strikes at random and cases are independent of one another.

What is the probability of 7 or more deaths from the disease on the British mainland (England, Scotland and Wales together) in any year?

*Solution*

Notice first that

(a)  P(7 or more deaths) = 1 − P(6 or fewer deaths);

(b)  Each of the three distributions fulfills the conditions for it to be modelled by the Poisson distribution.

You could find the probability of 6 or fewer deaths by listing all the different ways they could be allocated between the three countries (eg England 3, Scotland 1, Wales 2) and working out the probabilities of all of them, using the three individual Poisson distributions. That would be very time consuming indeed.

It is much easier to lump the three distributions together and treat the result as a single Poisson distribution.

The overall mean is given by     3.8     + 0.8     + 0.5   =  5.1

                         England   Scotland   Wales   Total

giving an overall distribution of Poisson(5.1).

The probability of 6 or less deaths is then found from Cumulative Poisson Probability tables to be 0.7474.

So the probability of 7 or more deaths is given by

$$1 - 0.7474 = 0.2526$$

**NOTES**

1.  *You may only add Poisson distributions in this way if they are independent of each other.*

2.  *The proof of the validity of adding Poisson distributions in this way is given in the Appendix on page 129.*

**EXAMPLE**

On a lonely Highland road in Scotland cars are observed passing at the rate of 6 per day, and lorries at the rate of 2 per day. On the road is an old cattle grid which will soon need repair. The local works department decide that if the probability of more than 15 vehicles per day passing is less than 1% then the repairs to the cattle grid can wait until next spring, otherwise it will have to be repaired before the winter.

When will the cattle grid have to be repaired?

*Solution*
Let $C$ be the number of cars per day, $L$ be the number of lorries per day, and $V$ be the number of vehicles per ~~hour~~. Day.

$$V = L + C$$

Assuming that a car or a lorry passing along the road is a random event and the two are independent:

$$C \sim \text{Poisson}(6), \quad L \sim \text{Poisson}(2)$$

and so  $V \sim \text{Poisson}(6 + 2)$

$$V \sim \text{Poisson}(8).$$

From Cumulative Poisson Probability tables

$$P(V \leq 15) = 0.9918$$

The required probability is  $P(V > 15) = 1 - P(V \leq 15)$

$$= 1 - 0.9918$$
$$= 0.0082.$$

This is just less than 1% and so the repairs can be left until spring.

## Modelling notes

The modelling of this situation raises a number of questions.

1. Is it true that a car or lorry passing along the road is a random event, or are some of these regular users, like the lorry collecting the milk from the farms along the road? If say 3 of the cars and one lorry are regular daily users, what effect does this have on the calculation?

2. Is it true that every car or lorry travels independently of every other one?

3. There are no figures for bicycles or motorbikes or other vehicles so it must be assumed that their numbers are negligible, or perhaps that they will cause no damage to the cattle grid.

1. Betty drives along a 50 mile stretch of motorway 5 days a week 50 weeks a year. She takes no notice of the 70 mph speed limit and, when the traffic allows, travels between 95 and 105 mph. From time to time she is caught by the police and fined but she estimates the probability of this happening on any day is 1/300. If she gets caught three times within three years she will be disqualified from driving. Use Betty's estimates of probability to answer the following questions.

   (i) What is the probability of her being caught exactly once in any year?
   (ii) What is the probability of her being caught less than three times in three years?
   (iii) What is the probability of her being caught exactly three times in three years?

   Betty is in fact caught one day and decides to be somewhat cautious, reducing her normal speed to between 85 and 95 mph. She believes this will reduce the probability of her being caught to 1/500.

   (iv) What is the probability that she is caught less than twice in the next three years?

2. Motorists in a particular part of the Highlands of Scotland have a choice between a direct route and a one-way scenic detour. It is known that on average one in forty of the cars on the road will take the scenic detour. The road engineer wishes to do some repairs on the scenic detour. He chooses a time when he expects 100 cars an hour to pass along the road.

   Find the probability that, in any one hour,

   (i) no cars;    (ii) at most 4 cars

   will turn on to the scenic detour.

   Between 10.30 am and 11.00 am it will be necessary to block the road completely. What is the probability that no car will be delayed?

3. A sociologist claims that only 3% of all suitably qualified students from inner city schools go on to university. Use his claim and the Poisson approximation to the Binomial distribution to estimate the probability that in a randomly chosen group of 200 such students

   i)   exactly five go to university
   ii)  more than five go to university.
   iii) If there is at most a 5% chance that more than $n$ of the 200 students go to university, find the lowest possible value of $n$.

   Another group of 100 students is also chosen. Find the probability that

   iv)  exactly five of each group go to university
   v)   exactly ten of all the chosen students go to university.          [MEI]

4. At the 'hot drinks' counter in a cafeteria both tea and coffee are sold. The number of cups of coffee sold per minute may be assumed to be a

MEI Structured Mathematics

*Exercise 2b continued*

Poisson variable with mean 1.5 and the number of cups of tea sold per minute may be assumed to be an independent Poisson variable with mean 0.5.

(i)   Calculate the probability that in a given one-minute period exactly one cup of tea and one cup of coffee are sold.

(ii)  Calculate the probability that in a given three-minute period fewer than 5 drinks altogether are sold.

(iii) In a given one-minute period exactly three drinks are sold. Calculate the probability that these are all cups of coffee.          [C]

5.  The numbers of lorry drivers and car drivers visiting an all-night transport cafe between 2 am and 3 am on a Sunday morning have independent Poisson distributions with means 5.1 and 3.6 respectively. Find the probabilities that, between 2 am and 3 am on any Sunday,

(i)   exactly 5 lorry drivers visit the cafe;

(ii)  at least one car driver visits the cafe;

(iii) exactly 5 lorry drivers and exactly 2 car drivers visit the cafe.

By using the distribution of the *total* number of drivers visiting the cafe, find the probability that exactly 7 drivers visit the cafe between 2 am and 3 am on any Sunday. Given that exactly 7 drivers visit the cafe between 2 am and 3 am on one Sunday, find the probability that exactly 5 of them are driving lorries.          [MEI]

6.  A garage uses a particular spare part at an average rate of 5 per week. Assuming that usage of this spare part follows a Poisson distribution, find the probability that

(i)   exactly 5 are used in a particular week,

(ii)  at least 5 are used in a particular week,

(iii) exactly 15 are used in a 3-week period,

(iv) at least 15 are used in a 3-week period,

(v)  exactly 5 are used in each of 3 successive weeks.

If stocks are replenished weekly, determine the number of spare parts which should be in stock at the beginning of each week to ensure that on average the stock will be insufficient on no more than one week in a 52 week year.          [AEB]

7.  Small hard particles are found in the molten glass from which glass bottles are made. On average 15 particles are found per 100 kg of molten glass. If a bottle contains one or more such particles it has to be discarded.

Suppose bottles of mass 1 kg are made. It is required to estimate the percentage of bottles that have to be discarded. Criticise the following 'answer': Since the material for 100 bottles contains 15 particles, approximately 15% will have to be discarded.

Making suitable assumptions, which should be stated, develop a correct argument using a Poisson model, and find the percentage of faulty 1 kg bottles to 3 significant figures.

Show that about 3.7% of bottles of mass 0.25 kg are faulty.                [MEI]

**8.** Weak spots occur at random in the manufacture of a certain cable at an average rate of 1 per 100 metres. If $X$ represents the number of weak spots in 100 metres of cable, write down the distribution of $X$.

Lengths of this cable are wound on to drums. Each drum carries 50 metres of cable. Find the probability that a drum will have 3 or more weak spots.

A contractor buys five such drums. Find the probability that two have just one weak spot each and the other three have none.

[AEB]

**9.** A crockery manufacturer tests dinner plates by taking a large sample from each day's production. When all the machinery is set correctly, there are on average 2 faulty plates per sample but this number rises if any part of the process is incorrectly set.

When the machine is working correctly, what is the probability that a test will result in:

(i)   no faulty plates;
(ii)  at most 4 faulty plates?

The company resets the machinery (a process which is expensive in time) if there are 5 or more faulty plates in a sample.

(iii) What is the probability that the machinery is reset unnecessarily?

The company decide to change the basis on which they make the decision to reset their machinery so that they will now do so if the total number of faulty plates in 3 consecutive samples is at least $f$.

(iv) Find the smallest value of $f$ which give a probability of less than 1% that the machinery will be reset unnecessarily.
(v)  Given your value of $f$, find the probability that a set of three consecutive samples fails to indicate that the machinery needs resetting when the mean number of faults per sample has in fact risen to 3.0.

**10.** A petrol station has service areas on both sides of a motorway, one to serve eastbound traffic and the other for westbound traffic. The number of eastbound vehicles arriving at the station in one minute has a Poisson distribution with mean 0.9, and the number of westbound

*Exercise 2b continued*

vehicles arriving in one minute has a Poisson distribution with mean 1.6, the two distributions being independent. Find the probability that in a one-minute period (*a*) no vehicles arrive, (*b*) more than two vehicles arrive at this petrol station, giving your answers correct to three places of decimals.

Given that in a particular one-minute period three vehicles arrive, find

(i) the probability that they are all from the same direction,

(ii) the most likely combination of eastbound and westbound vehicles.

[C]

**11.** The following table gives the number $f_r$ of each of 519 equal time intervals in which $r$ radioactive atoms decayed.

| Number of decays: $r$ | 0 | 1 | 2 | 3 | 4 | 5 | 6 | 7 | 8 | $\geq 9$ |
|---|---|---|---|---|---|---|---|---|---|---|
| Observed number of intervals: $f_r$ | 11 | 41 | 73 | 105 | 107 | 82 | 55 | 28 | 9 | 8 |

Estimate the mean and variance of $r$.

Suggest, with justification, a theoretical distribution from which the data could be a random sample. Hence calculate expected values of $f_r$ and comment briefly on the agreement between these and the observed values.

In the experiment each time interval was of length 7.5s. In a further experiment, 1000 time intervals each of length 5s are to be examined. Estimate the number of these intervals within which no atoms will decay.

# Summary

## The Poisson probability distribution

If $X \sim \text{Poisson}(\lambda)$          The parameter $\lambda > 0$.

$$P(X=r) = e^{-\lambda}\frac{\lambda^r}{r!} \qquad r \geq 0, r \text{ is an integer}$$

$$E(X) = \lambda$$

$$\text{Var}(X) = \lambda$$

## Conditions under which the Poisson distribution may be used

The Poisson distribution is generally thought of as the probability distribution for the number of occurrences of a *rare event*.

## As a distribution in its own right

Situations in which the mean number of occurrences is known (or can easily be found) but it is not possible, or even meaningful, to give values to $n$ or $p$ may be modelled using the Poisson distribution provided that the occurrences are

(i) random

(ii) independent.

## As an approximation to the binomial distribution

The Poisson distribution may be used as an approximation to the binomial distribution, B($n,p$), when

(i) $n$ is large

(ii) $p$ is small (and so the event is rare)

(iii) $np$ is not large.

It would be unusual to use the Poisson distribution with parameter, $\lambda$, greater than about 20.

# The sum of two Poisson distributions

If $X \sim$ Poisson $(\lambda)$, $Y \sim$ Poisson$(\mu)$ and $X, Y$ are independent $X + Y \sim$ Poisson $(\lambda + \mu)$

**HISTORICAL NOTE**

*Simeon Poisson was born in Pithiviers in France in 1781. Under family pressure he began to study medicine but after some time gave it up for his real interest, mathematics. For the rest of his life Poisson lived and worked as a mathematician in Paris. His contribution to the subject spanned a broad range of topics in both pure and applied mathematics, including integration, electricity and magnetism and planetary orbits as well as statistics. He was the author of between 300 and 400 publications and originally derived the Poisson distribution as an approximation to the binomial distribution.*

*When he was a small boy, Poisson had his hands tied by his nanny who then hung him from a hook on the wall so that he could not get into trouble while she went out. In later life he devoted a lot of time studying the motion of a pendulum, and claimed that this interest derived from his childhood experience, swinging against the wall.*

# 3  Normal distribution

*The
normal
law of error
stands out in the
experience of mankind
as one of the broadest
generalisations of natural
philosophy. It serves as the
guiding instrument in researches
in the physical and social sciences
and in medicine agriculture and engineering.
It is an indispensible tool for the analysis and the
interpretation of the basic data obtained by observation and experiment.*

W.J. Youden

---

THE AVONFORD STAR

## VILLAGERS GET GIANT BOBBY

The good people of Middle Fishbrook have special reason to be good these days. Since last week, their daily lives are being watched over by their new village bobby, Wilf 'Shorty' Harris.

At 6ft 4½in. Wilf is the tallest policeman in the county. "I don't expect any trouble", says Wilf. "But I wouldn't advise anyone to tangle with me on a dark night".

Seeing Wilf towering above me, I decided that most people would prefer not to put his words to the test.

*Towering Bobby 'Shorty' Harris is bound to deter mischief in Middle Fishbrook*

Wilf Harris is clearly exceptionally tall, but how much so? Is he one in a hundred, or a thousand or even million? To answer that question you need to know the distribution of the heights of adult British men.

Like many other naturally occurring variables, the heights of adult men may be modelled by the Normal distribution, shown below. You will see that this has a distinctive bell shaped curve, and is symmetrical about its middle. The curve is continuous; height is a continuous variable, not discrete.

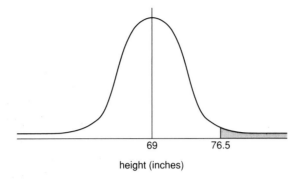

height (inches)

**Figure 3.1**

On figure 3.1, area represents probability so the shaded area to the right of 76.5in. represents the probability that a randomly selected adult male is over 6ft 4$\frac{1}{2}$in. (ie 76.5in) tall.

Before you can start to find this area, you must know the mean and standard deviation of the distribution, in this case about 5ft 9in. (ie 69in.) and 2.5in. respectively.

So Wilf's height is   76.5in. −69in.  =  7.5in.  above the mean, and that is

$$7.5 \div 2.5 \quad = \ 3 \quad \text{standard deviations.}$$

The number of standard deviations beyond the mean, in this case 3, is denoted by the letter $z$. Thus the shaded area gives the probability of obtaining a value of $z \geq 3$.

You find this area by looking up the value of $\Phi(z)$ when $z=3$ in a Normal distribution table of $\Phi(z)$ opposite, and then calculating $1 - \Phi(z)$. ($\Phi$ is the Greek letter phi, the nearest Greek equivalent to the letter F.)

This gives $\Phi(3) = 0.9987$, and so $1 - \Phi(3) = 0.0013$.

The probability of a randomly selected adult male being 6ft 4$\frac{1}{2}$in. or over is 0.0013.

Slightly more than one man in a thousand is at least as tall as Wilf.

# Using Normal distribution tables

The function $\Phi(z)$ gives the area under the Normal distribution curve to the left of the value $z$, that is the shaded area in figure 3.2 (It is the cumulative distribution function). The total area under the curve is 1, and the area given by $\Phi(z)$ represents the probability of a value smaller than $z$. Notice the scale for the $z$ values; it is in standard deviations from the mean.

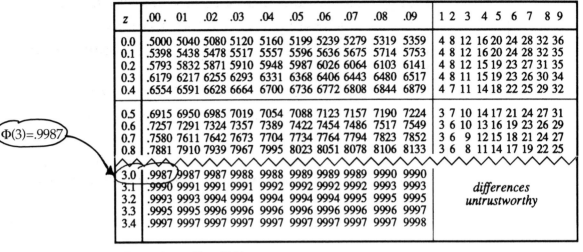

| z | .00 | .01 | .02 | .03 | .04 | .05 | .06 | .07 | .08 | .09 | 1 2 3 | 4 5 6 | 7 8 9 |
|---|---|---|---|---|---|---|---|---|---|---|---|---|---|
| 0.0 | .5000 | 5040 | 5080 | 5120 | 5160 | 5199 | 5239 | 5279 | 5319 | 5359 | 4 8 12 | 16 20 24 | 28 32 36 |
| 0.1 | .5398 | 5438 | 5478 | 5517 | 5557 | 5596 | 5636 | 5675 | 5714 | 5753 | 4 8 12 | 16 20 24 | 28 32 35 |
| 0.2 | .5793 | 5832 | 5871 | 5910 | 5948 | 5987 | 6026 | 6064 | 6103 | 6141 | 4 8 12 | 15 19 23 | 27 31 35 |
| 0.3 | .6179 | 6217 | 6255 | 6293 | 6331 | 6368 | 6406 | 6443 | 6480 | 6517 | 4 8 11 | 15 19 23 | 26 30 34 |
| 0.4 | .6554 | 6591 | 6628 | 6664 | 6700 | 6736 | 6772 | 6808 | 6844 | 6879 | 4 7 11 | 14 18 22 | 25 29 32 |
| 0.5 | .6915 | 6950 | 6985 | 7019 | 7054 | 7088 | 7123 | 7157 | 7190 | 7224 | 3 7 10 | 14 17 21 | 24 27 31 |
| 0.6 | .7257 | 7291 | 7324 | 7357 | 7389 | 7422 | 7454 | 7486 | 7517 | 7549 | 3 6 10 | 13 16 19 | 23 26 29 |
| 0.7 | .7580 | 7611 | 7642 | 7673 | 7704 | 7734 | 7764 | 7794 | 7823 | 7852 | 3 6 9 | 12 15 18 | 21 24 27 |
| 0.8 | .7881 | 7910 | 7939 | 7967 | 7995 | 8023 | 8051 | 8078 | 8106 | 8133 | 3 6 8 | 11 14 17 | 19 22 25 |
| 3.0 | .9987 | 9987 | 9987 | 9988 | 9988 | 9989 | 9989 | 9989 | 9990 | 9990 | | | |
| 3.1 | .9990 | 9991 | 9991 | 9991 | 9992 | 9992 | 9992 | 9992 | 9993 | 9993 | | *differences* | |
| 3.2 | .9993 | 9993 | 9994 | 9994 | 9994 | 9994 | 9994 | 9995 | 9995 | 9995 | | *untrustworthy* | |
| 3.3 | .9995 | 9995 | 9996 | 9996 | 9996 | 9996 | 9996 | 9996 | 9996 | 9997 | | | |
| 3.4 | .9997 | 9997 | 9997 | 9997 | 9997 | 9997 | 9997 | 9997 | 9997 | 9998 | | | |

$\Phi(3)=.9987$

*Extract from tables of $\Phi(z)$*

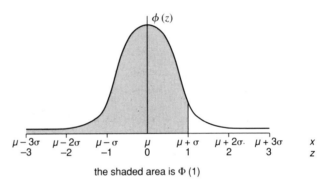

the shaded area is $\Phi(1)$

**Figure 3.2**

If the variable $X$ has mean $\mu$ and standard deviation $\sigma$, then $x$, a particular value of $X$, is transformed into $z$ by the equation

$$z = \frac{x-\mu}{\sigma}$$

$z$ is a particular value of the variable $Z$ which has mean 0 and standard deviation 1, and is the *standardised* form of the Normal distribution.

| | Actual distribution, $X$ | Standardised distribution, $Z$ |
|---|---|---|
| **Mean** | $\mu$ | 0 |
| **Standard deviation** | $\sigma$ | 1 |
| **Particular value** | $x$ | $z = \dfrac{x-\mu}{\sigma}$ |

Notice how lower case letters, $x$ and $z$, are used to indicate particular values of the random variables, whereas upper case letters, $X$ and $Z$, are used to describe or name those variables.

Normal distribution tables are easy to use but you should always make a point of drawing a diagram and shading the region you are interested in.

It is often helpful to know that in a Normal distribution, roughly:

68%, or about $\frac{2}{3}$, of the values lie within $\pm 1$ standard deviation of the mean;

95% of the values lie within $\pm 2$ standard deviations of the mean;

99.75% of the values lie within $\pm 3$ standard deviations of the mean.

**EXAMPLE**  Assuming the distribution of the heights of adult men is Normal, with mean 69in. and standard deviation 2.5in., find the probability that a randomly selected adult man is:

(i) under 6ft 1in.;  (ii) over 6ft 1in.;  (iii) over 5ft 11in.;  (iv) between 5ft 11in. and 6ft 1in.;  (v) under 5ft 7in.

giving answers to 2 significant figures,

*Solution*

The mean height,  $\mu = 69$

The standard deviation,  $\sigma = 2.5$

(i)  The probability that an adult man selected at random is under 6ft 1in.

The area required is that shaded in the diagram.

$x = 73$ (inches)

and so  $z = \dfrac{73 - 69}{2.5} = 1.6$

$\Phi(1.6) = 0.9452$
rounding to 0.95.

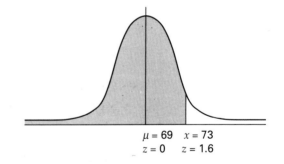

$\mu = 69 \quad x = 73$
$z = 0 \quad\ \ z = 1.6$

Answer: The probability that an adult man selected at random is under 6ft 1in. is 0.95.

(ii)  The probability that an adult man selected at random is over 6ft 1in.

The area required is the complement of that for part (i).

Probability $= 1 - \Phi(1.6)$

$= 1 - 0.9452$

$= 0.0548$
rounding to 0.055

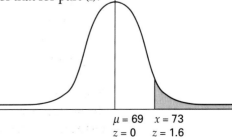

$\mu = 69 \quad x = 73$
$z = 0 \quad\ \ z = 1.6$

Answer: The probability that an adult man selected at random is over 6ft 1in. is 0.055.

(iii)  The probability that an adult man selected at random is over 5ft 11in.

$x = 71$ and so $z = \dfrac{71-69}{2.5} = 0.8$

The area required $= 1 - \Phi(0.8)$

$\qquad\qquad\quad = 1 - 0.7881$

$\qquad\qquad\quad = 0.2119$

$\qquad\qquad\qquad$ rounding to 0.21

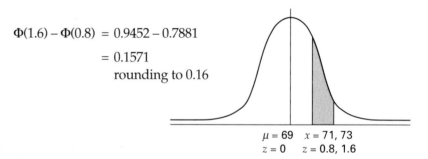

$\mu = 69 \quad x = 71$
$z = 0 \quad\;\; z = 0.8$

Answer: The probability that an adult man selected at random is over 5ft 11in. is 0.21.

(iv)  The probability that an adult man selected at random is between 5ft 11in. and 6ft 1in.

The required area is shown in the diagram.

It is

$\qquad\qquad \Phi(1.6) - \Phi(0.8) = 0.9452 - 0.7881$

$\qquad\qquad\qquad\qquad\qquad = 0.1571$

$\qquad\qquad\qquad\qquad\qquad$ rounding to 0.16

$\mu = 69 \quad x = 71, 73$
$z = 0 \quad\;\; z = 0.8, 1.6$

Answer: The probability that an adult man selected at random is over 5ft 11in. but under 6ft 1in. is 0.16.

(v)  The probability that an adult man selected at random is under 5ft 7in.

In this case $\qquad x = 67$

and so $\qquad\qquad z = \dfrac{67-69}{2.5} = -0.8$

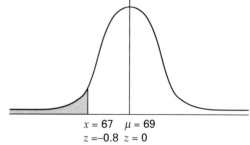

$x = 67 \quad \mu = 69$
$z = -0.8 \;\; z = 0$

However when you come to look up $\Phi(-0.8)$, you will find that only positive values of $z$ are given in your tables. You overcome this problem by using the symmetry of the Normal curve. The area you want in this case is that to the left of $-0.8$ and this is clearly just the same as that to the right of $+0.8$.

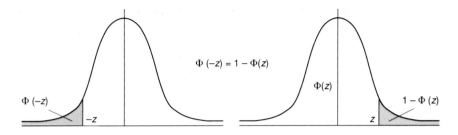

$$\Phi(-z) = 1 - \Phi(z)$$

$\Phi(-z)$    $\Phi(z)$    $1 - \Phi(z)$

$-z$    $z$

So    $\Phi(-0.8) = 1 - \Phi(0.8)$

$= 1 - 0.7881 = 0.2119$ rounding to 0.21.

Answer: The probability that an adult man selected at random is under 5ft 7in. is 0.21.

# The Normal curve

All Normal curves have the same basic shape, so that by scaling the two axes suitably you can always fit one normal curve exactly on top of another one.

The curve for the normal distribution with mean $\mu$ and standard deviation $\sigma$ (ie variance $\sigma^2$) is given by the function $\phi(x)$ in

$$\phi(x) = \frac{1}{\sigma\sqrt{2\pi}}\,e^{-\frac{1}{2}\left(\frac{x-\mu}{\sigma}\right)^2}$$

The notation $N(\mu,\sigma^2)$ is used to describe this distribution. The mean, $\mu$, and standard deviation, $\sigma$ (or variance, $\sigma^2$), are the two parameters used to define the distribution. Once you know their values, you know everything there is to know about the distribution. The standardised variable $Z$ has mean 0 and variance 1, so its distribution is $N(0,1)$.

After the variable $X$ has been transformed to $Z$ using $z = (x-\mu)/\sigma$ the form of the curve (now standardised) becomes

$$\phi(z) = \frac{1}{\sqrt{2\pi}}\,e^{-\frac{1}{2}z^2}$$

However, the exact shape of the Normal curve is often less useful than the area underneath it, which represents a probability.

The probability that $Z \le 2$ is given by the shaded area, and this is the shaded area in figure 3.3.

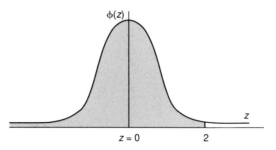

$\phi(z)$

$z = 0$     2     $z$

**Figure 3.3**

Easy though it looks, the function $\phi(z)$ cannot be integrated algebraically to find the area under the curve, only by a numerical method. The values found by doing so are given as a table and this area function is called $\Phi(z)$.

Make sure that you can distinguish between the upper and lower case Greek letter phi, $\Phi$ and $\phi$ respectively. You will find that you use $\phi(z)$ very little if at all, but often need the table of $\Phi(z)$.

**EXAMPLE**

Skilled operators make a particular component for an engine. The company believes that the time taken to make this component may be modelled by the Normal distribution with mean 95 minutes and standard deviation 4 minutes.

Assuming the company's belief to be true find the probability that the time taken to make one of these components, selected at random, was

(i) over 97 minutes; (ii) under 90 minutes; (iii) between 90 and 97 minutes.

Sheila believes that the company is allowing too long for the job and invites them to time her. They find that only 10% of the components take her over 90 minutes to make, and that 20% take her less than 70 minutes.

(iv) Estimate the mean and standard deviation of the time Sheila takes.

*Solution*

According to the company,

$$\mu = 95 \text{ and } \sigma = 4$$

The distribution is $N(95, 4^2)$.

(i)   The probability that a component requires over 97 minutes.

$$z = \frac{97 - 95}{4} = 0.5$$

The probability is represented by the shaded area and given by

$$1 - \Phi(0.5) = 1 - 0.6915$$
$$= 0.3085 \text{ rounding to } 0.31.$$

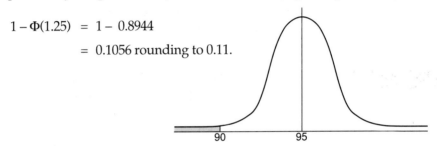

Answer: The probability that it took the operator over 97 minutes to manufacture a randomly selected component is 0.31.

(ii)   The probability that a component required under 90 minutes.

$$z = \frac{90 - 95}{4} = -1.25$$

The probability is represented by the shaded area and given by

$$1 - \Phi(1.25) = 1 - 0.8944$$
$$= 0.1056 \text{ rounding to } 0.11.$$

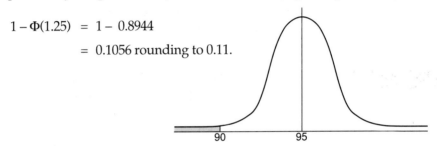

Answer: The probability that it took the operator under 90 minutes to manufacture a randomly selected component is 0.11.

(iii)   The probability that a component required between 90 and 97 minutes.

The probability is represented by the shaded area and given by

$$1 - 0.1056 - 0.3085 = 0.5859$$
rounding to 0.59.

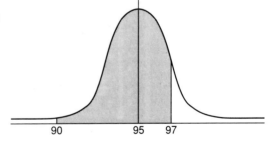

Answer: The probability that it took the operator between 90 and 97 minutes to manufacture a randomly selected component is 0.59.

(iv)   Estimate the mean and standard deviation of the time Sheila takes.

The question has now been put the other way round. You have to infer the mean, $\mu$, and standard deviation, $\sigma$, from the areas under different parts of the graph.

10% take her 90 minutes or more. This means that the shaded area is 0.1.

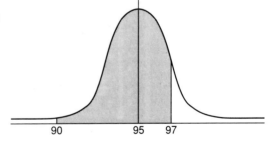

$$z = \frac{90 - \mu}{\sigma}$$

$$\Phi(z) = 1 - 0.1 = 0.9.$$

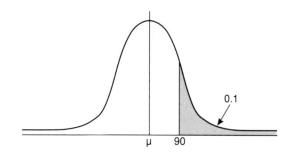

You must now use the table of the Inverse Normal function, $\Phi^{-1}(p) = z$

**The Inverse Normal function – values of $\Phi^{-1}(p) = z$**

$\Phi^{-1}(0.9)=1.282$

| $p$ | .000 | .001 | .002 | .003 | .004 | .005 | .006 | .007 | .008 | .009 |
|------|------|------|------|------|------|------|------|------|------|------|
| .50 | .0000 | .0025 | .0050 | .0075 | .0100 | .0125 | .0150 | .0175 | .0201 | .0226 |
| .51 | .0251 | .0276 | .0301 | .0326 | .0351 | .0376 | .0401 | .0426 | .0451 | .0476 |
| .52 | .0502 | .0527 | .0552 | .0577 | .0602 | .0627 | .0652 | .0677 | .0702 | .0728 |
| .89 | 1.227 | 1.232 | 1.237 | 1.243 | 1.248 | 1.254 | 1.259 | 1.265 | 1.270 | 1.276 |
| .90 | 1.282 | 1.287 | 1.293 | 1.299 | 1.305 | 1.311 | 1.317 | 1.323 | 1.329 | 1.335 |
| .91 | 1.341 | 1.347 | 1.353 | 1.360 | 1.366 | 1.372 | 1.379 | 1.385 | 1.392 | 1.398 |
| .92 | 1.405 | 1.412 | 1.419 | 1.426 | 1.433 | 1.440 | 1.447 | 1.454 | 1.461 | 1.468 |

*Extract from the Inverse Normal Distribution Table*

which tells you $z = 1.282$.

If you do not have the table of the Inverse Normal function, you can find the same answer by using the table of $\Phi(z)=p$ in reverse.

$\Phi^{-1}(0.9)=1.2815$

| $z$ | .00 | 01 | .02 | .03 | .04 | .05 | .06 | .07 | .08 | .09 | 1 2 3 | 4 5 6 | 7 8 9 |
|------|------|------|------|------|------|------|------|------|------|------|-------|-------|-------|
| 0.0 | .5000 | 5040 | 5080 | 5120 | 5160 | 5199 | 5239 | 5279 | 5319 | 5359 | 4 8 12 | 16 20 24 | 28 32 36 |
| 0.1 | .5398 | 5438 | 5478 | 5517 | 5557 | 5596 | 5636 | 5675 | 5714 | 5753 | 4 8 12 | 16 20 24 | 28 32 35 |
| 0.2 | .5793 | 5832 | 5871 | 5910 | 5948 | 5987 | 6026 | 6064 | 6103 | 6141 | 4 8 12 | 15 19 23 | 27 31 35 |
| 0.3 | .6179 | 6217 | 6255 | 6293 | 6331 | 6368 | 6406 | 6443 | 6480 | 6517 | 4 8 11 | 15 19 23 | 26 30 34 |
| 0.4 | .6554 | 6591 | 6628 | 6664 | 6700 | 6736 | 6772 | 6808 | 6844 | 6879 | 4 7 11 | 14 18 22 | 25 29 32 |
| 0.5 | .6915 | 6950 | 6985 | 7019 | 7054 | 7088 | 7123 | 7157 | 7190 | 7224 | 3 7 10 | 14 17 21 | 24 27 31 |
| 0.6 | .7257 | 7291 | 7324 | 7357 | 7389 | 7422 | 7454 | 7486 | 7517 | 7549 | 3 6 10 | 13 16 19 | 23 26 29 |
| 0.7 | .7580 | 7611 | 7642 | 7673 | 7704 | 7734 | 7764 | 7794 | 7823 | 7852 | 3 6 9 | 12 15 18 | 21 24 27 |
| 0.8 | .7881 | 7910 | 7939 | 7967 | 7995 | 8023 | 8051 | 8078 | 8106 | 8133 | 3 6 8 | 11 14 17 | 19 22 25 |
| 0.9 | .8159 | 8186 | 8212 | 8238 | 8264 | 8289 | 8315 | 8340 | 8365 | 8389 | 3 5 8 | 10 13 15 | 18 20 23 |
| 1.0 | .8413 | 8438 | 8461 | 8485 | 8508 | 8531 | 8554 | 8577 | 8599 | 8621 | 2 5 7 | 9 12 14 | 16 18 21 |
| 1.1 | .8643 | 8665 | 8686 | 8708 | 8729 | 8749 | 8770 | 8790 | 8810 | 8830 | 2 4 6 | 8 10 12 | 14 16 19 |
| 1.2 | .8849 | 8869 | 8888 | 8907 | 8925 | 8944 | 8962 | 8980 | 8997 | 9015 | 2 4 6 | 7 9 11 | 13 15 16 |
| 1.3 | .9032 | 9049 | 9066 | 9082 | 9099 | 9115 | 9131 | 9147 | 9162 | 9177 | 2 3 5 | 6 8 10 | 11 13 14 |
| 1.4 | .9192 | 9207 | 9222 | 9236 | 9251 | 9265 | 9279 | 9292 | 9306 | 9319 | 1 3 4 | 6 7 8 | 10 11 13 |

*Extract from tables of $\Phi(z)$*

Returning to the problem, you now know that

$$\frac{90 - \mu}{\sigma} = 1.282 \text{ simplifying to } 90 - \mu = 1.282\sigma.$$

The second piece of information, that 20% of components took her under 70 minutes, is illustrated in this diagram.

$$z = \frac{70 - \mu}{\sigma}$$

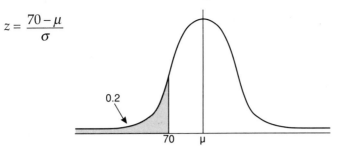

(Clearly $z$ has a negative value in this case, the point being to the left of the mean).

$\Phi(z) = 0.2$

and so, by symmetry, $\qquad \Phi(-z) = 1 - 0.2 = 0.8.$

Using the table of the Inverse Normal function gives

$$-z = 0.8416 \text{ or } z = -0.8416.$$

This gives a second equation for $\mu$ and $\sigma$.

$$\frac{70 - \mu}{\sigma} = -0.8416 \text{ simplifying to } 70 - \mu = -0.8416\sigma.$$

You now solve the two simultaneous equations

$$90 - \mu = 1.282\sigma$$
$$70 - \mu = -0.8416\sigma$$

Subtract $\quad 20 = 2.21236\sigma \quad \sigma = 9.418$ rounding to 9.4

and giving $\mu = 77.926$ rounding to 77.9.

Answer: Sheila's mean time is 77.9 minutes with standard deviation 9.4 minutes.

---

## Exercise 3A

1. The distribution of the heights of 18 year old girls may be modelled by the normal distribution with mean 162.5 cm and standard deviation 6 cm. Find the probability that the height of a randomly selected 18 year old girl is

   (i) under 168.5 cm; (ii) over 174.5 cm; (iii) between 168.5 and 174.5 cm.

2. A pet shop has a tank of goldfish for sale. All the fish in the tank were hatched at the same time and their weights may be taken to be normally

distributed with mean 100 g and standard deviation 10 g. Melanie is buying a goldfish and is invited to catch the one she wants in a small net. In fact the fish are much too quick for her to be able to catch any particular one and the fish which she eventually nets is selected at random. Find the probability that its weight is

(i) over 115 g; (ii) under 105 g; (iii) between 105 and 115 g.

**3.** When he makes instant coffee, Tony puts a spoonful of powder into a mug. The weight of coffee in grams in the spoon may be modelled by the normal distribution with mean 5 and standard deviation 1. If he uses more than 6.5 g Julia complains that it is too strong, and if he uses less than 4 g she tells him it is too weak. Find the probability that he makes the coffee

(i) too strong; (ii) too weak; (iii) all right.

**4.** When a butcher takes an order for a Christmas turkey, he asks the customer what weight in kg the bird should be. He then sends his order to a turkey farmer who supplies birds of about the requested weight. For any particular weight of bird ordered, their error in kg may be taken to be normally distributed with mean 0 and standard deviation 0.75.

Mrs. Jones orders a 10 kg turkey from the butcher. Find the probability that the one she gets is

(i) over 12 kg; (ii) under 10 kg; (iii) within 0.5 kg of the weight she actually ordered.

**5.** A biologist finds a nesting colony of a previously unknown sea bird on a remote island. She is able to take measurements on 100 of their eggs before replacing them in their nests. She records their weights, $w$ g, in this frequency table.

| Weight, $w$ | $25 < w \leq 27$ | $27 < w \leq 29$ | $29 < w \leq 31$ | $31 < w \leq 33$ | $33 < w \leq 35$ | $35 < w$ |
|---|---|---|---|---|---|---|
| **Frequency** | 2 | 13 | 35 | 33 | 17 | 0 |

(i) Use your calculator to find the mean and standard deviation of these data.

(ii) Assuming the weights of the eggs for this type of bird are Normally distributed, and that their mean and standard deviation are the same as those of this sample, find how many eggs you would expect to be in each of these categories.

(iii) Do you think the assumption that the weights of the eggs are Normally distributed is reasonable?

**6.** Hens' eggs have mean mass 60 g, with standard deviation 15 g, and the distribution may be taken as Normal. Eggs of mass less than 45 g are classified as Small. The remainder are divided into Standard and Large, and it is desired that these should occur with equal frequency. Suggest the mass at which the division should be made (correct to the nearest gram).

7. The length of life of a certain make of tyre is Normally distributed about a mean of 24 000 km with a standard deviation of 2500 km.

   (i) What percentage of such tyres will need replacing before they have travelled 20 000 km?
   (ii) As a result of improvements in manufacture, the length of life is still Normally distributed, but the proportion of tyres failing before 20 000 km is reduced to 1.5%.
   (A) If the standard deviation has remained unchanged, calculate the new mean length of life.
   (B) If, instead, the mean length of life has remained unchanged, calculate the new standard deviation. [MEI]

8. A machine is set to produce nails of lengths 10cm, with standard deviation 0.05 cm. The lengths of the nails are normally distributed.

   (i) Find the percentage of nails produced between 9.95 cm and 10.08 cm in length.

   The machine's setting is moved by a careless apprentice with the consequence that 16% of the nails are under 5.2 cm in length and 20% are over 5.3 cm.

   (ii) Find the new mean and standard deviation.

9. A marksman is firing at a target which consists of a pair of $x$ and $y$ axes graduated from $-30$ to $30$, where the units are centimetres. The bull is the origin.

   (i) The $x$-coordinates of his hits are distributed normally with a mean of zero. If, out of 100 shots, 34 lie within the range $-5 \leq x \leq 5$. find the standard deviation of this distribution.
   (ii) The $y$-coordinates of his hits are also distributed normally but have a mean of 4 and a standard deviation of 6. Find the probability of $y$ lying between $-5$ and $5$.
   (iii) The errors in the $x$-direction and the $y$-direction are unrelated. Find the probability that a hit will lie both in the region bounded by $x=-5$ and $x=5$, and also in the region bounded by $y=-5$ and $y=5$. [MEI]

10. The concentration by volume of methane at a point on the centre line of a jet of natural gas mixing with air is distributed approximately Normally with mean 20% and standard deviation 7%. Find the probabilities that the concentration

    (i) exceeds 30%;
    (ii) is between 5% and 15%.

Normal Distribution

*Exercise 3a continued*

In another similar jet, the mean concentration is 18% and the standard deviation is 5%. Find the probability that in at least one of the jets the concentration is between 5% and 15%. [MEI]

**11.** In a particular experiment, the length of a metal bar is measured many times. The measured values are distributed approximately Normally with mean 1.340 m and standard deviation 0.021 m. Find the probabilities that any one measured value

(i) exceeds 1.370 m;
(ii) lies between 1.310 m and 1.370 m;
(iii) lies between 1.330 m and 1.390 m.

Find the length $l$ for which the probability that any one measured value is less than $l$ is 0.1. [MEI]

**12.** Each weekday a man goes to work by bus. His arrival time at the bus stop is Normally distributed with standard deviation 3 minutes. His mean arrival time is 8.30 am. Buses leave promptly every 5 minutes at 8.21 am, 8.26 am, etc. Find the probabilities that he catches the buses at (i) 8.26 am; (ii) 8.31 am; (iii) 8.36 am, assuming that he always gets on the first bus to arrive.

The man is late for work if he catches a bus after 8.31 am. What mean arrival time would ensure that, on the average, he is not late for work more than one day in five? [Assume that he cannot change the standard deviation of his arrival time, and give your answer to the nearest 10s.] [MEI]

**13.** A machine is used to fill cans of soup with a nominal volume of 0.450 litres. Suppose that the machine delivers a quantity of soup which is Normally distributed with mean $\mu$ litres and standard deviation $\sigma$ litres. Given that $\mu = 0.457$ and $\sigma = 0.004$, find the probability that a randomly chosen can contains less than the nominal volume.

It is required by law that no more than 1% of cans contain less than the nominal volume. Find

(i) the least value of $\mu$ which will comply with the law when $\sigma = 0.004$;
(ii) the greatest value of $\sigma$ which will comply with the law when $\mu = 0.457$. [MEI]

**14.** A factory is lit by a large number of electric light bulbs whose lifetimes are modelled by a Normal distribution with mean 1000 hours and standard deviation 110 hours. Operating conditions require that all bulbs are on continuously. What proportions of bulbs have lifetimes that

(i) exceed 950 hours;
(ii) exceed 1050 hours?

Given that a bulb has already lasted 950 hours, what is the probability that it will last a further 100 hours? Give all answers correct to 3 decimal places.

The factory management is to adopt a policy whereby all bulbs will be replaced periodically after a fixed interval. To the nearest day, how long should this interval be if, on average 1% of the bulbs are to burn out between successive replacement times? [MEI]

**15.** A factory produces a very large number of steel bars. The lengths of these bars are normally distributed with 33% of them measuring 20.06 cm or more and 12% of them measuring 20.02 cm or less.

Write down two simultaneous equations for the mean and standard deviation of the distribution and solve to find values to 4 decimal places. Hence estimate the proportion of steel bars which measure 20.03 cm or more.

The bars are acceptable if they measure between 20.02 cm and 20.08 cm. What percentage are rejected as being outside the acceptable range? [MEI]

**16.** A characteristic of the shape of a human skull is measured by a number $n$. People are classed into three groups: $A$ (for which $n \leq 75$), $B$ ($75 < n \leq 80$) and $C$ ($n > 80$). In a certain population the percentages of people within these three groups are 58, 38 and 4 respectively. Assuming that $n$ is distributed Normally within this population, determine its mean and standard deviation.

Three people are chosen at random from this population. Determine the probabilities that

(i) each of the three has a value of $n$ greater than 70;
(ii) at least one of the three has a value of $n$ less than 70. [MEI]

**17.** The diameters $D$ of screws made in a factory are Normally distributed with mean 1 mm. Given that 10% of the screws have diameters greater than 1.04 mm, find the standard deviation correct to 3 significant figures, and hence show that about 2.7% of the screws have diameters greater than 1.06 mm.

Find, correct to 2 significant figures,

(i) the number $d$ for which 99% of the screws have diameters that exceed $d$ mm,
(ii) the number $\varepsilon$ for which 99% of the screws have diameters that do not differ from the mean by more than $\varepsilon$ mm. [MEI]

**18.** A firm makes two different brands of similar electronic components, $A$ and $B$. The life of brand $A$ has mean 23 hours, standard deviation 2 hours; the life of brand $B$ has mean 25 hours, standard deviation 5 hours. Their lives are assumed to be distributed according to a Normal probability model. Which brand is more likely to break down over a period of
(i) 22 hours, (ii) 20 hours?

*Exercise 3a continued*

Replacing such a component in the course of a certain job causes expensive delay, so a new component is fitted before starting the job. If the component lasts for $x$ hours, it ensures a profit £$F(x,)$ where

$$F(x) = \begin{cases} -100, & 0 < x < 20; \\ 20, 20 \le x < 25; \\ 40, & x \ge 25. \end{cases}$$

Which brand gives the greater expected profit? [SMP]

**19.** A machine produces crankshafts whose diameters are Normally distributed with mean 5 cm and standard deviation 0.03 cm. Find the percentage of crankshafts it will produce whose diameters lie between 4.95 cm and 4.97 cm.

What is the probability that two successive crankshafts will both have a diameter in this interval?

Crankshafts with diameters outside the interval $5 \pm 0.05$ cm are rejected. If the mean diameter of the machine's production remains unchanged, to what must the standard deviation be reduced if only 4% of the production is to be rejected? [MEI]

**20.** An aircraft component has a life before failure whose mean is 20 months and standard deviation 4 months.

(a) The component is meant to be replaced on an aircraft so that the probability of it failing in service is 0.1 or less. After how many months should one of these components be replaced?

(b) One of these components is not replaced when it should be. Calculate the probability that it will last 26 months or more before failure.

(c) The component is now redesigned so that the standard deviation of its life is changed (but the mean kept the same). The component need now only be replaced after 18 months. Calculate the new standard deviation. [MEI]

# The use of the Normal distribution to model discrete situations

Although the Normal distribution applies strictly to a continuous variable, it is also common to use it in situations where the variable is discrete providing that:

● the distribution is approximately normal; this requires that the steps in its possible values are small compared with its standard deviation;

● *continuity corrections* are applied where appropriate.

The meaning of the term continuity correction is explained in the following example.

The result of an Intelligence Quotient (IQ) test is an integer score, $X$. Tests are designed so that $X$ has a mean value of 100 with standard deviation 15. A large number of people have their IQs tested. What proportion of them would you expect to have IQs measured between 106 and 110 (inclusive)?

*Solution*

Although the random variable $X$ is an integer and hence discrete, the steps of 1 in its possible values are small compared with the standard deviation of 15. So it is reasonable to treat it as if it is continuous.

If you assume that an IQ test is measuring innate, natural intelligence (rather than the results of learning) then it is reasonable to assume a Normal distribution.

If you draw the probability distribution function for the discrete variable $X$ it looks like Figure 3.4. The area you require is the total of the 5 bars representing 106, 107, 108, 109 and 110.

The equivalent section of the normal curve would run not from 106 to 110 but from 105.5 to 110.5, as you can see in the diagram. When you change from the discrete scale to the continuous scale, the numbers 106, 107 etc no longer represent the whole intervals, just their centre points.

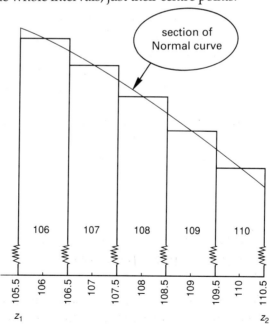

**Figure 3.4**

So the area you require under the Normal curve is given by

$$\Phi(z_2) - \Phi(z_1)$$

Normal Distribution

where $z_1 = \dfrac{105.5 - 100}{15}$ and $z_2 = \dfrac{110.5 - 100}{15}$

This is $\Phi(0.7000) - \Phi(0.3667)$

$= 0.7580 - 0.6431$

$= 0.1149$

Answer: The proportion of IQs between 106 and 110 (inclusive) should be approximately 11%.

In this calculation, both end values needed to be adjusted to allow for the fact that a continuous distribution was being used to approximate a discrete. These adjustments, $106 \rightarrow 105.5$ and $110 \rightarrow 110.5$, are called continuity corrections. Whenever a discrete distribution is approximated by a continuous one a continuity correction may need to be used.

You must always think carefully when applying a continuity correction. Should the corrections be added or subtracted? In this case 106 and 110 are inside the required area and so any value (like 105.7 or 110.4) which would round to them must be included. It is often helpful to draw a sketch to illustrate the region you want, like the one above.

If the region of interest is given in terms of inequalities, you should look carefully whether they are inclusive ($\leq$ or $\geq$) or exclusive ($<$ or $>$). For example

$20 \leq X \leq 30$ becomes $19.5 \leq X \leq 30.5$

whereas $20 < X < 30$ becomes $20.5 \leq X \leq 29.5$

Two particularly common situations are when the Normal distribution is used to approximate the binomial and the Poisson distributions.

# The Normal distribution as an approximation for the binomial distribution

You may use the Normal distribution as an approximation for the binomial, $B(n,p)$ (where $n$ is the number of trials each having probability $p$ of success) when

1. $n$ is large;
2. $p$ is not too close to 0 or 1.

These conditions ensure that the distribution is reasonably symmetrical and not skewed away from either end, figure 3.5.

The parameters for the Normal distribution are then

Mean: $\mu = np$

Variance: $\sigma^2 = npq$

so that it can be denoted by $N(np, npq)$.

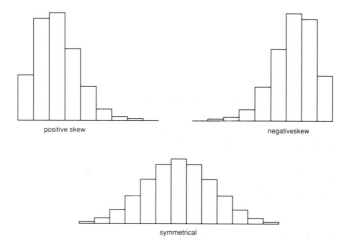

positive skew

negativeskew

symmetrical

**Figure 3.5**

(This is a true story).

During voting at a by-election, an exit poll of 1700 voters indicated that 50% of people had voted for the Labour party candidate. When the real votes were counted it was found that he had in fact received 57% support.

850 of the 1700 people interviewed said they had voted Labour but 57% of 1700 is 969, a much higher number. What went wrong? Is it possible to be so far out just by being unlucky and asking the wrong people?

The situation of selecting a sample of 1700 people and asking them if they voted for one party or not is one that is modelled by the binomial distribution, in this case B(1700,0.57).

In theory you could multiply out $(0.43 + 0.57t)^{1700}$ and use that to find the probabilities of getting 0, 1, 2, ..., 850 Labour voters in your sample of 1700. In practice of course such a method would be completely impractical because of the work involved.

What you can do is to use a normal approximation. The required conditions are fulfilled: at 1700, $n$ is certainly not small; $p = 0.57$ is near neither 0 nor 1.

The parameters for the Normal approximation are given by

$$\text{Mean, } \mu \quad = \quad np \quad = \quad 1700 \times 0.57 \quad = 969$$
$$\text{S.D., } \sigma \quad = \sqrt{npq} \quad = \sqrt{1700 \times 0.57 \times 0.43} \quad = 20.4$$

You will see that the standard deviation, 20.4, is large compared to the steps of 1 in the possible values of Labour voters.

The probability of getting no more than 850 Labour voters, P(X≤850), is given by $\Phi(z)$, where

$$z = \frac{850.5 - 969}{20.4} = -5.8.$$

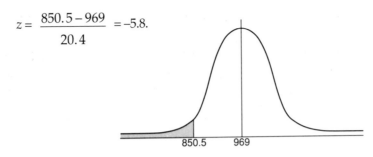

850.5     969

(Notice the continuity correction making 850 into 850.5).

This is beyond the range of most tables and corresponds to a probability of about 0.000 01. The probability of a result as extreme as this is thus 0.000 02 (allowing for an equivalent result in the tail above the mean). It is clearly so unlikely that this was a result of random sampling that another explanation must be found.

So what went wrong with the exit poll?

One possibility is that some people, knowing their votes should be secret, resented being asked who they had supported and so deliberately gave wrong answers. Another is that the exit poll was taken at a time of day when those voting were unrepresentative of the electorate as a whole.

What do you think? Remember this really did happen.

**NOTE**   *This situation was presented in* Statistics 1 *(page 112) for readers to investigate using a computer simulation.*

# The Normal distribution as an approximation for the Poisson distribution

You may use the Normal distribution as an approximation for the Poisson distribution, provided that its parameter (mean) $\lambda$ is sufficiently large for the distribution to be reasonably symmetrical and not positively skewed (to the right).

As a working rule $\lambda$ should be at least 10. That gives

$$\text{mean} \quad = 10$$

and standard deviation $= \sqrt{10} = 3.16$.

A Normal distribution is almost entirely contained within 3 standard deviations of its mean, and in this case the value 0 is slightly more than 3 standard deviations away from the mean value of 10.

The parameters for the Normal distribution are then

$$\text{Mean:} \quad \mu = \lambda$$
$$\text{Variance:} \quad \sigma^2 = \lambda$$

so that it can be denoted by $N(\lambda, \lambda)$.

(Remember that for a Poisson distribution, Mean = Variance).

**EXAMPLE** The annual number of deaths nationally from a rare disease, $X$, may be modelled by the Poisson distribution with mean 25. One year there are 31 deaths and it is suggested that the disease is on the increase.

What is the probability of 31 or more deaths in a year, assuming the mean has remained at 25?

*Solution*

The Poisson distribution with mean 25 may be approximated by the Normal distribution with parameters

Mean:   25

S.D.:   $\sqrt{25} = 5$

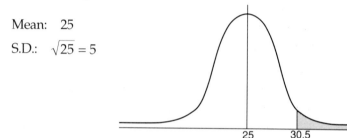

The probability of there being 31 or more deaths in a year, $P(X \geq 31)$, is given by $1 - \Phi(z)$, where

$$z = \frac{30.5 - 25}{5} = 1.1$$

(Note the continuity correction, replacing 31 by 30.5).

The required area is $1 - \Phi(1.1)$

$$= 1 - 0.8643$$
$$= 0.1357.$$

This is not a particularly low probability; it is quite likely that there would be that many deaths in any one year.

## Exercise 3B

**1.** The intelligence of an individual is frequently described by a positive integer known as an IQ (intelligence quotient). The distribution of IQs amongst children of a certain age-group can be approximated by a Normal probability model with mean 100 and standard deviation 15. Write a sentence stating what you understand about the age-group from the fact that $\Phi(2.5) = 0.994$.

A class of 30 children is selected at random from the age-group. Calculate (to 2 significant figures) the probability that at least one member of the class has an IQ of 138 or more.                   [SMP]

**2.** A certain examination has a mean mark of 100, and a standard deviation of 15. The marks can be assumed to be Normally distributed.

(i) What is the least mark needed to be in the top 35% of pupils taking this examination?

(ii) Between which two marks will the middle 90% of the pupils lie?

(iii) 150 pupils take this examination. Calculate the number of pupils likely to score 110 or over. [MEI]

**3.** 25% of Flapper Fish have red spots, the rest blue spots. A fisherman nets 10 Flapper Fish. What are the probabilities that

(i) exactly 8 have blue spots;

(ii) at least 8 have blue spots.

A large number of samples, each of 100 Flapper Fish, are taken.

(iii) Where are  (A)  the mean;

(B)  the standard deviation of the number of red spotted fish per sample?

(iv) What is the probability of a sample of 100 Flapper Fish containing over 30 with red spots?

**4.** A fair coin is tossed 10 times. Evaluate the probability that exactly half of the tosses result in heads.

The same coin is tossed 100 times. Use the Normal approximation to the Binomial to estimate the probability that exactly half of the tosses result in heads. Also estimate the probability that more than 60 of the tosses result in heads.

Explain briefly the meaning of the term 'continuity correction' in using the Normal approximation to the Binomial, and the reason for the adoption of this correction. [MEI]

**5.** In what circumstances may the Normal distribution be used as an approximation to the Binomial distribution with $n$ trials and a probability $p$ of success in each trial?

Ten per cent of the screws manufactured by a machine are defective.

Find the mean and standard deviation of the number of defective screws in a batch of 1000. Hence find an approximate upper limit $L$ such that 99% of these batches have fewer than $L$ defective screws. [O&C]

**6.** During an advertising campaign, the manufacturers of Wolfitt, (a dog food) claimed that 60% of dog owners preferred to buy Wolfitt. Assuming that the manufacturer's claim is correct for the population of dog owners, calculate

(a) using the Binomial distribution, and

(b) using a Normal approximation to the Binomial,

the probability that at least 6 of a random sample of 8 dog owners prefer to buy Wolfitt. Comment on the agreement, or disagreement, between your two values. Would the agreement be better or worse if the proportion had been 80% instead of 60%?

Continuing to assume that the manufacturer's figure of 60% is correct, use the Normal approximation to the Binomial to estimate the probability that, of a random sample of 100 dog owners, the number preferring Wolfitt is between 60 and 70 inclusive. [MEI]

**7.** A multiple-choice examination consists of 20 questions, for each of which the candidate is required to tick as correct one of 3 possible answers. Exactly one answer to each question is correct. A correct answer gets 1 mark and a wrong answer gets 0 marks. Consider a candidate who has complete ignorance about every question and therefore ticks at random. What is the probability that he gets a particular answer correct? Calculate the mean and variance of the number of questions he answers correctly.

The examiners wish to ensure that not more than 1% of completely ignorant candidates pass the examination. Use the Normal approximation to the Binomial, working throughout to 3 decimal places, to establish the pass mark that meets this requirement. [MEI]

**8.** State the precise conditions under which the Normal distribution can be used as an approximation to the Binomial distribution.

A machine produces screws, some of which are known to be faulty. An inspection scheme is devised by taking random samples of 1000 screws from a large batch produced by the machine and noting the number $x$ of faulty screws in the sample. If $x > 80$ the batch is rejected, but if $x \le 80$ the batch is accepted. Find the probabilities of

(i) accepting a batch containing 5% faulty screws,
(ii) rejecting a batch containing 10% faulty screws. [MEI]

**9.** An inter-city telephone exchange has 100 lines and on average 80 are in use at any moment (on a typical business-day morning). Calculate

(i) the probability that all lines are engaged;
(ii) the probability that more than 30 lines are free.

We say that a number $x$ of lines is the 'effective minimum level' if the number of lines in use exceeds $x$ for 95% of the time. Find $x$. [SMP]

**10.** A telephone exchange serves 2000 subscribers, and at any moment during the busiest period there is a probability of $\frac{1}{30}$ for each subscriber that he will require a line. Assuming that the needs of subscribers are independent, write down an expression for the probability that exactly $N$ lines will be occupied at any moment during the busiest period.

Use the Normal distribution to estimate the minimum number of lines that would ensure that the probability that a call cannot be made because all the lines are occupied is less than $0.01$.

Investigate whether the total number of lines needed would be reduced if the subscribers were split into two groups of 1000, each with its own set of lines. [MEI]

**11.** The number of cars per minute entering a multi-storey car park can be modelled by a Poisson distribution with mean 2. What is the probability that 3 cars enter during a period of one minute?

What are the mean and the standard deviation of the number of cars entering the car park during a period of 30 minutes? Use the Normal approximation to the Poisson distribution to estimate the probability that at least 50 cars enter in any one 30 minute period. [MEI]

**12.** Define the Poisson distribution. State its mean and variance. State under what circumstances the Normal distribution can be used as an approximation to the Poisson distribution.

Readings, on a counter, of the number of particles emitted from a radioactive source in a time $T$ seconds have a Poisson distribution with mean $250\,T$. A ten-second count is made. Find the probabilities of readings of (i) more than 2600, (ii) 2400 or more. [JMB]

**13.** The quantity of milk in bottles from a dairy is Normally distributed with mean $1.036$ pints and standard deviation $0.014$ pints.

Show that the probability of a randomly chosen bottle containing less than a pint is very nearly $0.5\%$.

In the rest of this question take the answer of $0.5\%$ to be exact. A crate contains 24 bottles. Find the probability that

(i)  no bottles contain less than a pint of milk,
(ii) at most 1 bottle contains less than a pint of milk.

A milk float is loaded with 150 crates (3600 bottles) of milk. State the expected number of bottles containing less than a pint of milk.

Give a suitable approximating distribution for the number of bottles containing less than a pint of milk. Use this distribution to find the probability that more than 20 bottles contain less than a pint of milk. [MEI]

**14.** In the country of Statia the mean number of deaths each year from lightning strikes is 2. Assuming that the deaths are independent, find the probability that the number of deaths from lightning strikes next year will be is

i) 0;   ii) 1;   iii) 2.

In the neighbouring country of Proba there are no deaths from lightning strikes just one year in twenty.

*Exercise 3b continued*

    iv) Calculate the mean number of deaths per year in Proba from lightning strikes. Give your answer to 3 decimal places and use it to this accuracy in the next two parts.

    v) Find the probability that next year a total of exactly 5 people will be killed by lightning strikes in the two countries.

    vi) Use the Normal approximation to the Poisson distribution to find the probability that in a century more than 480 people in total will be killed by lightning strikes in the two countries.

**15.** A large computer system which is in constant operation requires an average of 30 service calls per year.

    (i) State the average number of service calls per month, taking a month to be $\frac{1}{12}$ of a year. What assumptions need to be made for the Poisson distribution to be used to model the number of calls in a given month?

    (ii) Use the Poisson distribution to find the probability that at least one service call is required in January. Obtain the probability that there is at least one service call in each month of the year.

    (iii) The service contract offers a discount if the number of service calls in the year is 24 or fewer. Use a suitable approximating distribution to find the probability of obtaining the discount in any particular year.

                                                   [MEI]

# The Central Limit Theorem

The Normal distribution occupies a central position in statistics. In this chapter you have met it in situations where it occurs naturally and also as an approximation to other distributions. Its use, however, goes well beyond this.

Statistical work often involves drawing conclusions from samples. In order to do so you need to understand the distributions of sample statistics, and in particular the sample mean. This is described by the Central Limit Theorem:

*For samples of size n drawn from a distribution with mean $\mu$ and finite variance $\sigma^2$, the distribution of the sample mean is approximately $N(\mu, \sigma^2/n)$ for sufficiently large n.*

Even if the underlying distribution is not Normal, the distribution of the means of samples of a particular size drawn from it is approximately Normal. The larger the sample size, the closer is this distribution to the Normal.

So, even if the Normal distribution never occurred in nature you would still need to know about it when drawing conclusions from sample data.

If the underlying distribution is normal, then the distribution of the sample mean is Normal whatever the size of $n$.

*Experiment*

The distribution of the scores on a die is uniform, probability $\frac{1}{6}$ for each of the outcomes 1, 2, 3, 4, 5 and 6. It is not a Normal Distribution.

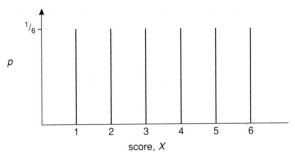

Throw five dice (or one die 5 times) and record the total score. This will be somewhere between 5 (5 ones) and 30 (5 sixes) and probably somewhere in the middle of the range. If it were, say, 16, then the sample mean would be $^{16}/_5 = 3.2$. Mark this on a graph.

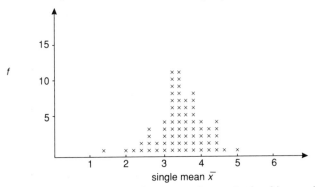

Repeat the experiment many times, and watch the Normal curve grow on your graph.

This experiment can obviously be conducted with different numbers of dice. The resulting mean will actually be a discrete variable (with steps of 0.2 in this case); the larger the number of dice the more accurate will be the ultimate curve, but the longer the experiment will take you. It can also be set up as a computer simulation.

# Normal probability graph paper

There will be times when you will want to judge whether data you have collected could have come from a Normal distribution. There are formal tests for establishing this, but they are outside the scope of this book. A simple procedure is to use *Normal probability graph paper*, as in the following example.

**EXAMPLE**

The weights of 50 adult female pet cats were measured as follows.

| Weight (kg) | $2.0<w\leq3.0$ | $3.0<w\leq4.0$ | $4.0<w\leq5.0$ | $5.0<w\leq6.0$ | $6.0<w\leq7.0$ |
|---|---|---|---|---|---|
| Frequency | 2 | 9 | 15 | 12 | 6 |

| Weight (kg) | $7.0<w\leq8.0$ | $8.0<w\leq9.0$ | $9.0<w\leq10.0$ |
|---|---|---|---|
| Frequency | 3 | 1 | 2 |

Do these data appear to be a sample from a Normal distribution?

*Solution*

The first step is to construct a cumulative frequency table and then to write the frequencies as percentage probabilities.

| Weight (kg) | Frequency | Percentage |
|:---:|:---:|:---:|
| ≤2 | 0 | 0 |
| ≤3 | 2 | 4 |
| ≤4 | 11 | 22 |
| ≤5 | 26 | 52 |
| ≤6 | 38 | 76 |
| ≤7 | 44 | 88 |
| ≤8 | 47 | 94 |
| ≤9 | 48 | 96 |
| ≤10 | 50 | 100 |

The percentage figures are then plotted on Normal probability graph paper, figure 3.6, but without the first and last point (0% and 100%). If a straight line results, they would indeed seem to be a sample from a Normal distribution.

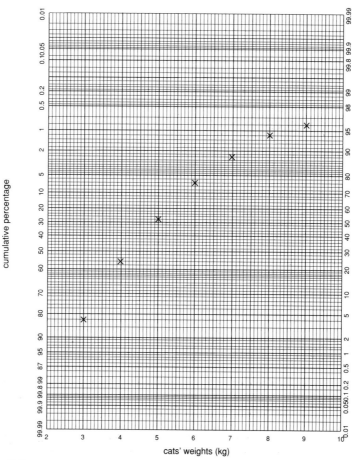

**Figure 3.6**

You can see that these data do not give a straight line. It seems that the distribution of the weights of female pet cats is not normal. In fact if you look at the figures in 3.7 you will see at a glance that there is considerable positive skew, presumably due to a number of very heavy cats, possibly fat creatures, spending their lives curled up by the fire!

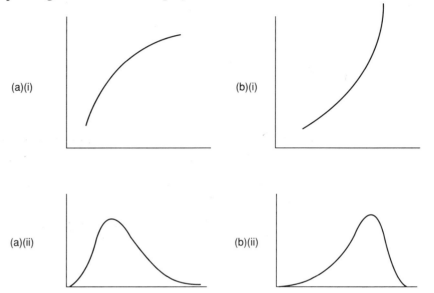

(a)(i)

(b)(i)

(a)(ii)

(b)(ii)

**Figure 3.7 (a) Positive Skew, (b) Negative Skew. Drawn: (i) on Normal probability graph paper, (ii) as a distribution**

The shape of this graph indicates that the data has positive skew.

## Summary

The Normal distribution with mean $\mu$ and standard deviation $\sigma$ is denoted by $N(\mu,\sigma^2)$.

This may be given in standardised form by using the transformation

$$z = \frac{x - \mu}{\sigma}$$

In the standardised form, $N(0,1)$, the mean is 0, the standard deviation and variance both 1.

The standardised normal curve is given by

$$\Phi(z) = \frac{1}{\sqrt{2\pi}}\, e^{-\frac{1}{2}z^2}$$

The area to the left of the value $z$, in figure 3.8, representing the probability of a value less than $z$, is denoted by $\Phi(z)$ and is read from tables.

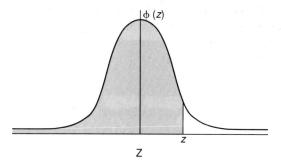

**Figure 3.8**

The normal distribution may be used to approximate suitable discrete distributions but continuity corrections are then required.

The binomial distribution B($n,p$) may be approximated by N($np,npq$), provided $n$ is large and $p$ is not close to 0 or 1.

The Poisson distribution Poisson($\lambda$) may be approximated by N($\lambda,\lambda$), provided $\lambda$ is about 10 or more.

# Bivariate data

*It is now proved beyond doubt that smoking is one of the leading causes of statistics.*

John Peers

---

THE AVONFORD STAR

# Traffic Related Death

**Reporter Ruth Williams gives an in-depth analysis of the statistics that mean pain and suffering to families the length and breadth of the country.**

Last week a motorist was fined £2500 for knocking down a young man on one of Avonford's pedestrian crossings. Having seen the grief of the victim's family, my first reaction was that £25,000 and a few years in prison would have been more appropriate, but that was before I met the motorist. Nothing that a court could do to him would ever take away the memory of the worst split second of his life.

"One careless moment and my life is ruined", he told me, "I am not a killer. All I was trying to do was to keep up with the traffic on a murky evening. You know what it's like, cars and lorries everywhere, and you are trying to sort out the thousand and one lights coming at you. But then suddenly out of nowhere there was this face right in front of me. It will haunt me for as long as I live".

I don't blame him now but say "There but for the grace of God go I".

*The crossing in Main Street, Avonford, where a motorist hit and killed a young man last month*

It is statistically inevitable that the more vehicles we cram into our roads, the more human tragedies will occur. Just look at the figures for some other countries and you will see what I mean.

*Continued overleaf*

| Country | Vehicles/1000 population | Road deaths/10,000 population |
|---|---|---|
| Greece | 150 | 1.7 |
| Spain | 295 | 1.8 |
| Belgium | 377 | 2.0 |

The conclusion is inescapable: the greater the traffic density, the greater the number of deaths.

The data in this example are a set of pairs of values for two variables, the traffic density and the death rate from traffic accidents. This is an example of *bivariate data*, where each point requires the values of two variables. The graph in the article is called a *scatter diagram* and this is a common way of showing bivariate data.

If you can draw a straight line through the data, then there is said to be perfect *linear correlation* between the two variables. It is much more likely however that your data fall close to a straight line but not exactly on it. The better the fit, the higher the level of linear correlation.

The term *line of best fit* is used to describe a straight line drawn through a set of data points so as to fit them as closely as possible. There are several ways of determining such a line, according to what precisely is meant by a close fit.

## Discussion

Do you agree with the conclusion in the article? Do you accept the argument that the reporter uses, or do you think it is over-simplified?

Before making your judgement look at the data for all the E.C. countries, from which those three were extracted, and plot the points on a scatter diagram. What has the reporter done to make her case seem stronger?

| Country | Vehicles/1000 population | Road deaths/10 000 population |
|---|---|---|
| Belgium | 377 | 2.0 |
| Denmark | 311 | 1.3 |
| France | 410 | 1.9 |
| Germany | 479 | 1.3 |
| Greece | 150 | 1.7 |
| Rep. of Ireland | 222 | 1.3 |
| Italy | 425 | 1.1 |
| Luxemburg | 470 | 1.7 |
| Netherlands | 351 | 1.0 |
| Portugal | 227 | 2.3 |
| Spain | 295 | 1.8 |
| U.K. | 366 | 1.0 |

Source: *The Independent* 23/9/92

# Interpreting Scatter Diagrams

You can often judge if correlation is present just by looking at a scatter diagram, see figure 4.1.

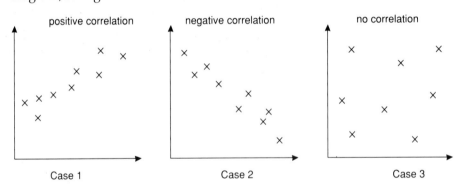

**Figure 4.1**

In case 1 both the variables increase together.

In case 2 as one variable increases the other variable decreases.

In case 3 as one variable increases there is no clear pattern as to how the other variable behaves.

## Investigations

There are many situations you can investigate for yourself. Here are two which you can carry out with a group of your friends.

## Dice

Toss a pair of ordinary but distinguishable dice, A and B, 100 times and record the following data for each toss:

*(side margin)* MEI Structured Mathematics

(a) the score on die A;

(b) the total score on the two dice;

(c) the difference between the scores on each die.

Plot 3 scatter diagrams to compare (a) and (b); (a) and (c); (b) and (c). Are you able to draw any conclusions about the relationship between the data?

## Brains

Select a group of students of about the same age.

For each student, find the circumference of his or her head and the total time spent doing homework in the last week. Plot the data on a scatter diagram. Do you think there is a hint of correlation present?

# Line of best fit

If your scatter diagram leads you to suspect that there is linear correlation between the two variables plotted then you may reasonably try to draw a line of best fit. A simple, but not very accurate, way to do this is as follows.

(a) Calculate and plot the point $(\bar{x}, \bar{y})$ where $\bar{x}$ is the mean of the horizontal axis variable and $\bar{y}$ is the mean of the vertical axis variable.

(b) Draw a straight line which passes through $(\bar{x}, \bar{y})$ and which roughly leaves the same number of points of the scatter diagram above and below it, see figure 4.2.

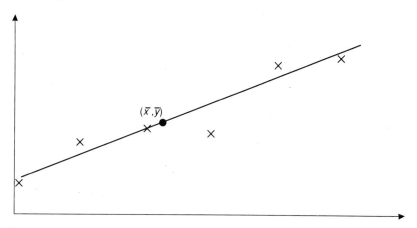

**Figure 4.2**

In some situations this rough and ready method will be adequate, particularly where it is easy to see where the line should go. In others cases, you may find it hard to judge where to place the line and may well want to draw the line accurately anyway. The method for doing this is described later in this chapter on pages 00 to 00.

# Dependent and independent variables

In the article about the traffic accidents, the scatter diagram was drawn with the traffic density on the horizontal axis and the death rate on the vertical axis. It was done that way because the reporter wanted to show that the death rate was dependent on the traffic density. This is normal practice, to plot the *dependent variable* on the vertical axis and the *independent variable* on the horizontal axis. In some cases the independent variable is non-random, like the data below for the memory quotient of a rat (on a particular scale), and its age.

| Age (months) | 6 | 12 | 18 | 24 | 30 |
|---|---|---|---|---|---|
| Memory quotient | 16 | 22 | 25 | 25 | 23 |

Quite clearly the memory is dependent in some way on the age of the rat but its age, being just the length of time since it was born, is independent of its memory quotient. In this example the age is also a non-random variable since the tests were clearly conducted every six months and there was nothing random about when they would take place; it is a controlled variable.

During the later part of this chapter you will be asked to consider the nature of the variables you are dealing with in more detail.

## Exercise 4A

For each of the sets of data below, draw a scatter diagram and comment on whether there appears to be any correlation. If there is then draw a possible line of best fit.

**1.** The Mathematics and Physics test results of 14 pupils.

| $X$: **Mathematics** | 45 | 23 | 78 | 91 | 46 | 27 | 41 | 62 | 34 | 17 | 77 | 49 | 55 | 71 |
|---|---|---|---|---|---|---|---|---|---|---|---|---|---|---|
| $Y$: **Physics** | 62 | 36 | 92 | 70 | 67 | 39 | 61 | 40 | 55 | 33 | 65 | 59 | 35 | 40 |

**2.** The wine consumption in a country in millions of litres and the years 1983 to 1990.

| Year | 1983 | 1984 | 1985 | 1986 | 1987 | 1988 | 1989 | 1990 |
|---|---|---|---|---|---|---|---|---|
| Consumption ($\times 10^6$ litres.) | 35.5 | 37.7 | 41.5 | 46.4 | 44.8 | 45.8 | 53.9 | 62.0 |

**3.** The monthly rainfall in cm, and the number of hours of sunshine in an 8 month period.

| | JAN | FEB | MAR | APR | MAY | JUN | JUL | AUG |
|---|---|---|---|---|---|---|---|---|
| **Rainfall (cm)** | 5.1 | 4.6 | 6.3 | 5.1 | 3.3 | 2.8 | 4.5 | 4.0 |
| **Sunshine (hrs)** | 90 | 96 | 105 | 110 | 113 | 120 | 131 | 124 |

4. The annual salary, in thousands of pounds, and the average number of hours worked per week by 7 people chosen at random.

| Salary ($\times$ £1000) | 5 | 7 | 13 | 14 | 16 | 20 | 48 |
|---|---|---|---|---|---|---|---|
| Hours worked per week | 18 | 22 | 35 | 38 | 36 | 36 | 32 |

5. The mean temperature in degrees centigrade and the amount of ice-cream sold in a supermarket in hundreds of litres.

|  | APR | MAY | JUN | JUL | AUG | SEP | OCT | NOV |
|---|---|---|---|---|---|---|---|---|
| Mean temperature ($^\circ$C) | 9 | 13 | 14 | 17 | 16 | 15 | 13 | 11 |
| Ice-cream sold (100 litres) | 11 | 15 | 17 | 20 | 22 | 17 | 8 | 7 |

6. The reaction times of 10 women of various ages

| Reaction time ($\times 10^{-3}$s) | 156 | 165 | 149 | 180 | 189 | 207 | 208 | 178 |
|---|---|---|---|---|---|---|---|---|
| Age (years) | | 36 | 40 | 27 | 50 | 49 | 53 | 55 | 27 |

# Product Moment Correlation

The football league table below shows the position of teams in the English Premier Division towards the middle of the 1992–3 season.

**Teams and Tables**

**FA Premier League**

| Pos | | P | Home W | D | L | F | A | Away W | D | L | F | A | Pts |
|---|---|---|---|---|---|---|---|---|---|---|---|---|---|
| 1 | Norwich | 17 | 6 | 2 | 0 | 13 | 6 | 5 | 1 | 3 | 19 | 24 | 36 |
| 2 | Blackburn | 17 | 6 | 1 | 2 | 18 | 7 | 2 | 6 | 0 | 8 | 5 | 31 |
| 3 | Arsenal | 17 | 6 | 0 | 3 | 14 | 8 | 3 | 2 | 3 | 8 | 9 | 29 |
| 4 | Aston Villa | 17 | 4 | 3 | 2 | 15 | 10 | 3 | 4 | 1 | 11 | 8 | 28 |
| 5 | Man Utd | 17 | 3 | 3 | 2 | 9 | 7 | 4 | 3 | 2 | 9 | 5 | 27 |
| 6 | Q.P.R | 17 | 4 | 3 | 1 | 15 | 9 | 3 | 2 | 4 | 7 | 8 | 26 |
| 7 | Man City | 17 | 3 | 3 | 3 | 14 | 10 | 4 | 1 | 3 | 10 | 7 | 25 |
| 8 | Liverpool | 17 | 6 | 1 | 2 | 21 | 10 | 1 | 3 | 4 | 9 | 14 | 25 |
| 9 | Chelsea | 16 | 3 | 3 | 2 | 10 | 8 | 4 | 1 | 3 | 13 | 11 | 25 |
| 10 | Ipswich | 17 | 3 | 6 | 0 | 13 | 9 | 2 | 4 | 2 | 9 | 10 | 25 |
| 11 | Coventry | 17 | 2 | 2 | 4 | 9 | 12 | 4 | 4 | 1 | 12 | 10 | 24 |
| 12 | Tottenham | 17 | 3 | 4 | 1 | 11 | 8 | 2 | 3 | 4 | 6 | 14 | 22 |
| 13 | Leeds | 16 | 5 | 3 | 0 | 20 | 7 | 0 | 3 | 5 | 8 | 20 | 21 |
| 14 | Middlesbro | 17 | 4 | 3 | 1 | 16 | 8 | 1 | 3 | 5 | 11 | 19 | 21 |
| 15 | Sheff Wed | 17 | 3 | 3 | 2 | 11 | 10 | 1 | 5 | 3 | 8 | 10 | 20 |
| 16 | Southmptn | 17 | 2 | 4 | 2 | 8 | 8 | 2 | 3 | 4 | 7 | 11 | 19 |
| 17 | Oldham | 17 | 4 | 3 | 2 | 20 | 14 | 0 | 3 | 5 | 7 | 16 | 18 |
| 18 | Sheff Utd | 17 | 3 | 5 | 1 | 10 | 8 | 1 | 1 | 6 | 7 | 15 | 18 |
| 19 | Everton | 17 | 1 | 3 | 4 | 4 | 10 | 3 | 1 | 5 | 9 | 11 | 16 |
| 20 | Wimbledon | 17 | 1 | 3 | 5 | 9 | 14 | 2 | 3 | 3 | 10 | 12 | 15 |
| 21 | Crystal Pal | 17 | 0 | 5 | 3 | 10 | 13 | 1 | 4 | 4 | 10 | 19 | 12 |
| 22 | Nottm For | 17 | 2 | 1 | 6 | 5 | 10 | 0 | 4 | 4 | 8 | 17 | 11 |

home goals / points /

Is there any correlation between the number of goals scored at home by the various teams and their total numbers of points?

The first step you would take in looking at this situation is to draw a scatter diagram of the relevant data, figure 4.3.

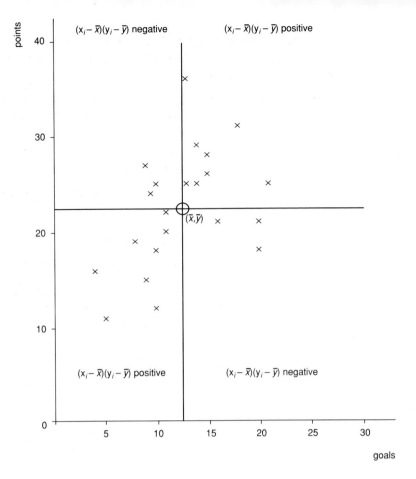

**Figure 4.3**

The mean number of home goals is $\bar{x} = \Sigma x_i/n = 275/22 = 12.5$.

The mean number of points is $\bar{y} = \Sigma y_i/n = 494/22 = 22.45$.

[$(x_i, y_i)$ are the various data points, like $(18,31)$ for Blackburn Rovers. $n$ is the number of such points, in this case 22, one for each club in the Premier League.] You will see that the point $(\bar{x},\bar{y})$ has also been plotted on the scatter diagram and lines drawn through this point parallel to the axes. These lines divide the scatter diagram up into four regions.

You can think of the point $(\bar{x},\bar{y})$ as the middle of the scatter diagram and so treat it as a new origin. Relative to $(\bar{x},\bar{y})$, the co-ordinates of the various points are all of the form $(x_i{-}\bar{x}, y_i{-}\bar{y})$

In regions 1 and 3 the product $(x_i{-}\bar{x})(y_i{-}\bar{y})$ is positive for every point.

In regions 2 and 4 the product $(x_i{-}\bar{x})(y_i{-}\bar{y})$ is negative for every point.

When there is positive correlation most or all of the data points will fall in regions 1 and 3 and so you would expect the sum of these terms,

$$\sum_i (x_i - \bar{x})(y_i - \bar{y})$$

to be positive and large.

When there is negative correlation most or all of the points will be in regions 2 and 4 and so you would expect the sum of these terms, (in the equation above) to be negative and large.

When there is little or no correlation the points will be scattered round all four regions. Those in regions 1 and 3 will result in positive values of $(x_i-\bar{x})(y_i-\bar{y})$ but when you add these to the negative values from the points in regions 2 and 4 you would expect most of them to cancel each other out. Consequently the value of the terms should be small.

By itself the actual value of $\sum_i(x_i-\bar{x})(y_i-\bar{y})$ does not tell you very much because:

1. no allowance has been made for the number of items of data;

2. no allowance has been made for the spread within the data;

3. no allowance has been made for the units of $x$ and $y$.

Taking the first point into account converts $\sum_i(x_i-\bar{x})(y_i-\bar{y})$ into *sample covariance*, $s_{xy}$. If allowance is made for the second and third points as well the *product moment correlation coefficient* of the sample is found.

## 1. Covariance

The first point is met by finding the average value of $\sum_i(x_i-\bar{x})(y_i-\bar{y})$. This is called the *covariance* and given by

$$\text{Covariance} = s_{xy} = \frac{1}{n}\sum_i(x_i - \bar{x})(y_i - \bar{y})$$

where $n$ is the number of data points.

Covariance may also be written in the form $\dfrac{1}{n}\sum_i(x_iy_i)-\bar{x}\bar{y}.$

You can easily show that the two forms are equivalent, as follows.

$$\frac{1}{n}\sum_i(x_i - \bar{x})(y_i - \bar{y}) = \frac{1}{n}\sum_i(x_iy_i - x_i\bar{y} - \bar{x}y_i + \bar{x}\bar{y})$$

$$= \frac{1}{n}\sum_i x_iy_i - \bar{y}\times\frac{1}{n}\sum_i x_i - \bar{x}\times\frac{1}{n}\sum_i y_i + \frac{1}{n}\sum_i \bar{x}\bar{y}$$

Remember $\dfrac{1}{n}\sum_i x_i = \bar{x},\ \dfrac{1}{n}\sum_i y_i = \bar{y}$    Since $\bar{x}$ and $\bar{y}$ are constant

$$\sum_i \bar{x}\bar{y} = n\bar{x}\bar{y}$$

$$= \frac{1}{n}\sum_i x_iy_i - \bar{y}\bar{x} - \bar{x}\bar{y} + \frac{1}{n}\times n\bar{x}\bar{y}$$

$$= \frac{1}{n}\sum_i x_iy_i - \bar{x}\bar{y}$$

This allows for the number of data points but not for the spread of the variables.

## 2. Pearson's Product Moment Correlation Coefficient

To allow for the spread of the data, you must standardise the values of $(x_i - \bar{x})$ and $(y_i - \bar{y})$ by dividing by their standard deviations $s_x$ and $s_y$, given by:

$$s_x = \sqrt{\frac{1}{n}\Sigma(x_i - \bar{x})^2} \qquad \text{and} \qquad s_y = \sqrt{\frac{1}{n}\Sigma(y_i - \bar{y})^2}$$

The resulting measure of linear correlation is called *Pearson's Product Moment Correlation Coefficient* and is denoted by the letter $r$.

$$r = \frac{s_{xy}}{s_x s_y}$$

$$r = \frac{\dfrac{1}{n}\Sigma(x_i - \bar{x})(y_i - \bar{y})}{\sqrt{\dfrac{1}{n}\Sigma(x_i - \bar{x})^2}\sqrt{\dfrac{1}{n}\Sigma(y_i - \bar{y})^2}}$$

which can be simplified to

$$r = \frac{\Sigma(x_i - \bar{x})(y_i - \bar{y})}{\sqrt{\Sigma(x_i - \bar{x})^2 \Sigma(y_i - \bar{y})^2}}$$

An alternative form for this is

$$r = \frac{\dfrac{1}{n}\Sigma x_i y_i - \bar{x}\bar{y}}{\sqrt{\left(\dfrac{1}{n}\Sigma x_i^2 - \bar{x}^2\right)\left(\dfrac{1}{n}\Sigma y_i^2 - \bar{y}^2\right)}}$$

$r$ is often just called the correlation coefficient.

The quantity, $r$, provides a standardised measure of correlation.

Its value always lies within the range –1 to +1. (If you calculate a value outside this range, you have made a mistake). A value of +1 means perfect positive correlation; in that case all the points on a scatter diagram would lie exactly on a straight line with positive gradient. Similarly a value of –1 perfect negative correlation, see figure 4.4.

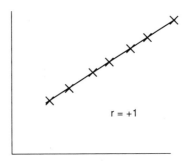

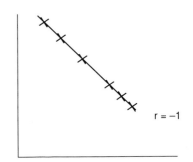

**Figure 4.4 (i) Perfect positive correlation**    **(ii) Perfect negative correlation**

In cases of little or no correlation $r$ takes values close to zero. The nearer the value of r is to +1 or –1, the stronger the correlation.

The calculation of covariance and Pearson's product moment correlation coefficient is shown in the following example, which is worked twice, using the alternative formulae.

HISTORICAL NOTE

*Karl Pearson was one of the founders of modern statistics. Born in 1857, he was a man of varied interests and practiced law for three years before being appointed Professor of Applied Mathematics and Mechanics at University College, London in 1884. Pearson made contributions to various branches of mathematics but is particularly remembered for his work on the application of statistics to biological problems in heredity and evolution. He died in 1936.*

EXAMPLE

A gardener wishes to know if plants which produce only a few potatoes also produce large ones. He selects 5 plants at random, sieves out the small potatoes, counts those remaining and weighs the largest one.

| Number of potatoes ($x$) | 5 | 6 | 7 | 8 | 9 |
|---|---|---|---|---|---|
| Weight of largest ($y$ oz) | 9.6 | 9.3 | 9.1 | 8.9 | 8.6 |

Calculate the covariance and correlation coefficient between $x$ and $y$.

*Solution* (Method 1)

| $x_i$ | $y_i$ | $x_i-\bar{x}$ | $y_i-\bar{y}$ | $(x_i-\bar{x})^2$ | $(y_i-\bar{y})^2$ | $(x_i-\bar{x})(y_i-\bar{y})$ | |
|---|---|---|---|---|---|---|---|
| 5 | 9.6 | –2 | 0.5 | 4 | 0.25 | –1.0 | |
| 6 | 9.3 | –1 | 0.2 | 1 | 0.04 | –0.2 | $n = 5$ |
| 7 | 9.1 | 0 | 0 | 0 | 0 | 0 | |
| 8 | 8.9 | 1 | –0.2 | 1 | 0.04 | –0.2 | $\bar{x} = 7$ |
| 9 | 8.6 | 2 | –0.5 | 4 | 0.25 | –1.0 | |
| Σ   35 | 45.5 | 0 | 0 | 10 | 0.58 | –2.4 | $\bar{y} = 9.1$ |

$$\bar{x} = \frac{\Sigma x_i}{n} = \frac{35}{5} = 7 \qquad \bar{y} = \frac{\Sigma y_i}{n} = \frac{45.5}{5} = 9.1$$

$$s_x = \sqrt{\frac{1}{n}\Sigma(x_i - \bar{x})^2} = \sqrt{(10/5)} = 1.414$$

$$s_y = \sqrt{\frac{1}{n}\Sigma(y_i - \bar{y})^2} = \sqrt{(0.58/5)} = 0.341$$

$$s_{xy} = \frac{1}{n}\Sigma(x_i - \bar{x})(y_i - \bar{y}) = \frac{-2.4}{5} = -0.48 \; Covariance$$

$$r = \frac{s_{xy}}{s_x s_y} = \frac{-0.48}{1.414 \times 0.341} = -0.995 \; Correlation \; coefficient$$

Bivariate data

91 | S2

*Solution* (Method 2)

| | $x_i$ | $y_i$ | $x_i^2$ | $y_i^2$ | $x_iy_i$ | |
|---|---|---|---|---|---|---|
| | 5 | 9.6 | 25 | 92.16 | 48.0 | |
| | 6 | 9.3 | 36 | 86.49 | 55.8 | |
| | 7 | 9.1 | 49 | 82.81 | 63.7 | $n = 5$ |
| | 8 | 8.9 | 64 | 79.21 | 71.2 | |
| | 9 | 8.6 | 81 | 73.96 | 77.4 | |
| Totals   $\Sigma$ | 35 | 45.5 | 255 | 414.63 | 316.1 | |

$$\bar{x} = \frac{\Sigma x_i}{n} = \frac{35}{5} = 7 \qquad \bar{y} = \frac{\Sigma y_i}{n} = \frac{45.5}{5} = 9.1$$

$$s_x = \sqrt{\left(\frac{1}{n}\Sigma(x_i^2 - \bar{x}^2)\right)} = \sqrt{\left(\frac{255}{5} - 7^2\right)} \qquad = 1.414$$

$$s_y = \sqrt{\left(\frac{1}{n}\Sigma y_i^2 - \bar{y}^2\right)} = \sqrt{\left(\frac{414.63}{5} - 9.1^2\right)} \qquad = 0.341$$

$$s_{xy} = \frac{1}{n}\Sigma x_iy_i - \bar{x}\bar{y} \quad = \frac{316.1}{5} - 7 \times 9.1 \qquad = -0.48 \quad \textit{Covariance}$$

$$r = \frac{s_{xy}}{s_xs_y} \qquad = \frac{-0.48}{1.414 \times 0.341} \qquad = -0.995 \; \textit{Correlation coefficient}$$

There is very strong negative linear correlation between the variables. Large potatoes seem to be associated with small crop sizes.

---

## Exercise 4B

In this exercise you are asked to find the covariance $s_{xy}$ and Pearson's product moment correlation coefficient, $r$, for a number of bivariate samples. The purpose of doing this on paper is to familiarise yourself with the routines involved. Most statisticians would actually do such calculations using a computer package, spreadsheet or a good calculator, and you should learn how to do that as well.

**1.**

| $x$ | 2 | 6 | 7 | 10 |
|---|---|---|---|---|
| $y$ | 13 | 8 | 9 | 6 |

**2.**

| $x$ | 10 | 11 | 12 | 13 | 14 | 15 | 16 | 17 |
|---|---|---|---|---|---|---|---|---|
| $y$ | 19 | 16 | 28 | 20 | 31 | 19 | 32 | 35 |

**3.**

| $x$ | 0 | 1 | 4 | 3 | 2 |
|---|---|---|---|---|---|
| $y$ | 11 | 8 | 5 | 4 | 7 |

**4.**

| $x$ | 12 | 14 | 14 | 15 | 16 | 17 | 17 | 19 |
|---|---|---|---|---|---|---|---|---|
| $y$ | 86 | 90 | 78 | 71 | 77 | 69 | 80 | 73 |

**5.**

| $x$ | 56 | 78 | 14 | 80 | 34 | 78 | 23 | 61 |
|---|---|---|---|---|---|---|---|---|
| $y$ | 45 | 34 | 67 | 70 | 42 | 18 | 25 | 50 |

# The meaning of a correlation coefficient

You have already seen that if the value of the correlation coefficient, $r$, is close to +1 or –1, you can be satisfied that there is linear correlation, and that if $r$ is close to 0 there is probably little or no correlation. What happens in a case such as $r = 0.6$?

To answer that question you have to understand what $r$ is actually measuring. The data which you use when calculating $r$ are actually a *sample* of a parent bivariate distribution. You have only taken a few out of a very large number of points which could, in theory, be plotted on a scatter diagram, figure 4.5. Each point $(x_i, y_i)$ represents one possible value $x_i$ of the variable $X$ and one possible value $y_i$ of $Y$.

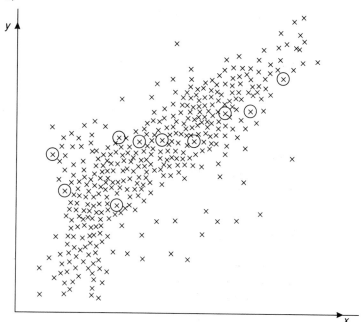

**Figure 4.5 Scatter diagram showing a sample from a large bivariate population**

There will be a level of correlation within the parent population and this is denoted by $\rho$ (rho, the Greek letter r, pronounced like 'row', as in 'row a boat').

Your calculated value of $r$, which is based on your sample points, can be used as an estimate for $\rho$. It can also be used to carry out a hypothesis test on the value of $\rho$, the parent population correlation coefficient. Used in this way it is a *test statistic*.

The simplest hypothesis test which you can carry out is that there is no correlation within the parent population. This gives rise to a null hypothesis:

$\mathrm{H}_0 \colon \rho = 0$  There is no correlation between the two variables.

There are three possible alternative hypotheses, according to the sense of the situation you are investigating. These are:

$H_1$:    $\rho \neq 0$    There is correlation between the variables
(2-tail test)

or

$H_1$:    $\rho > 0$    There is positive correlation between the variables
(1-tail test)

or

$H_1$:    $\rho < 0$    There is negative correlation between the variables
(1-tail test)

The test is carried out by comparing your value for $r$ with the appropriate entry in a table of critical values. This will depend on the size of your sample, the significance level at which you are testing and whether your test is 1– or 2–tailed.

**EXAMPLE**

THE AVONFORD STAR

# Letters to the Editor

Dear Sir,

The trouble with young people these days is that they watch too much television. They just sit there gawping and become steadily less intelligent. I challenge you to carry out a proper test and I am sure you will find that the more telly children watch the less intelligent they are.

Yours truthfully,

Outraged Senior citizen.

The editor of the *Avonford Star* was interested in the writer's point of view and managed to collect these data for the IQs of 6 children, and the number of hours of television they watched in the previous week.

| Hours of TV ($x$): | 9 | 11 | 14 | 7 | 10 | 9 |
|---|---|---|---|---|---|---|
| IQ ($y$): | 142 | 112 | 100 | 126 | 109 | 88 |

The relevant hypothesis test for this situation would be

$H_0$:    $\rho = 0$    There is no correlation between IQ and watching television

$H_1$:    $\rho < 0$    There is negative correlation between IQ and watching television.
(1-tail test).

The editor decided to use 5% significance level.

The critical value for $n=6$ at the 5% significance level for a 1-tail test is found from tables to be 0.7293.

| 5% | 2½% | 1% | ½% | 1–Tail Test |
|---|---|---|---|---|
| 10% | 5% | 2% | 1% | 2–Tail Test |

| $n$ | | | | |
|---|---|---|---|---|
| 1 | – | – | – | – |
| 2 | – | – | – | – |
| 3 | 0.9877 | 0.9969 | 0.9995 | 0.9999 |
| 4 | 0.9000 | 0.9500 | 0.9800 | 0.9900 |
| 5 | 0.8054 | 0.8783 | 0.9343 | 0.9587 |
| 6 | 0.7293 | 0.8114 | 0.8822 | 0.9172 |
| 7 | 0.6694 | 0.7545 | 0.8329 | 0.8745 |
| 8 | 0.6215 | 0.7067 | 0.7887 | 0.8343 |
| 9 | 0.5822 | 0.6664 | 0.7498 | 0.7977 |
| 10 | 0.5494 | 0.6319 | 0.7155 | 0.7646 |
| 11 | 0.5214 | 0.6021 | 0.6851 | 0.7348 |
| 12 | 0.4973 | 0.5760 | 0.6581 | 0.7079 |
| 13 | 0.4762 | 0.5529 | 0.6339 | 0.6835 |
| 14 | 0.4575 | 0.5324 | 0.6120 | 0.6614 |
| 15 | 0.4409 | 0.5140 | 0.5923 | 0.6411 |

*Extract from table of values for the product moment correlation coefficient, r*

The calculation of the correlation coefficient can be set out as follows. (Notice that most of the figures are given to 2 decimal places, but greater accuracy is used in the actual calculation).

| $x_i$ | $y_i$ | $x_i^2$ | $y_i^2$ | $x_i y_i$ |
|---|---|---|---|---|
| 9 | 142 | 81 | 20164 | 1278 |
| 11 | 112 | 121 | 12544 | 1232 |
| 14 | 100 | 196 | 10000 | 1400 |
| 7 | 126 | 49 | 15876 | 882 |
| 10 | 109 | 100 | 11881 | 1090 |
| 9 | 88 | 81 | 7744 | 792 |
| $\Sigma$  60 | 677 | 628 | 78209 | 6674 |

$$\bar{x} = \frac{\Sigma x_i}{n} = \frac{60}{6} = 10 \qquad \bar{y} = \frac{\Sigma y_i}{n} = \frac{677}{6} = 112.83$$

$$s_x = \sqrt{\left(\frac{1}{n}\Sigma x_i^2 - \bar{x}^2\right)} = \sqrt{\left(\frac{628}{6} - 10^2\right)} \qquad = 2.16$$

$$s_y = \sqrt{\left(\frac{1}{n}\Sigma y_i^2 - \bar{y}^2\right)} = \sqrt{\left(\frac{78209}{6} - 112.83^2\right)} = 17.42$$

$$s_{xy} = \frac{1}{n}\Sigma x_i y_i - \bar{x}\bar{y} = \frac{6674}{6} - 10 \times 112.83 = -16.00 \text{ Covariance}$$

$$r = \frac{s_{xy}}{s_x s_y} = \frac{-16.00}{2.16 \times 17.42} = -0.43 \text{ Correlation coefficient}$$

Since 0.43 < 0.7293, the critical value, the null hypothesis is accepted.

The evidence from this small sample is not sufficient to justify the claim that there is negative correlation between IQ and television watching. Notice however that even if it had been, this would not necessarily have supported *'Outraged Senior Citizen's'* claim that television lowers IQ. It could just be that people with higher IQs watch less television.

**N O T E**

# Degrees of freedom

*The extract from the tables gives the critical value of* **r** *for various values of the significance level and the sample size, n. You will however find some tables where n is replaced by* $\nu$*, the degrees of freedom. (*$\nu$*, is the Greek letter n and is pronounced like 'new'.*

Here is an example where you have just 2 data points.

|  | Sean | Iain |
|---|---|---|
| **Height of an adult man** (m): | 1.70 | 1.90 |
| **The mortgage on his house** (£): | 15 000 | 45 000 |

When you plot these two points on a scatter diagram it is possible to join them with a perfect straight line and you might be tempted to conclude that 'Taller men have larger mortgages on their houses'.

That conclusion would clearly be wrong. It is based on the data from only two men so you are bound to be able to join the points on the scatter diagram with a straight line and calculate $r$ to be either +1 or −1 (providing their heights and/or mortgages are not the same). In order to start to carry out a test you need the data for a third man, say Dafyd (height 1.75 m and mortgage £37 000). When his data are plotted on the scatter diagram you can see how close it lies to the line between Sean and Iain, see figure 4.6.

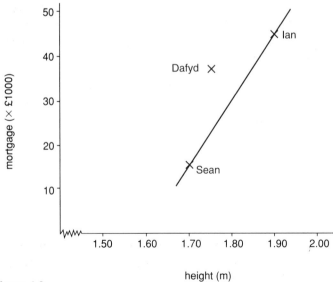

**Figure 4.6**

So the first two data points do not count towards a test for linear correlation. The first one to count is point number three. Similarly if you have $n$ points, only $n-2$ of them count towards any test. $n-2$ is called the *degrees of freedom* and denoted by $v$. It is the number of free variables with the system, the number of points, in this case $n$, less the 2 that have effectively been used to define the line of best fit.

In the case of the three men with their mortgages you would actually draw a line of best fit through all three, rather than join any particular two. So you cannot say that any two particular points have been taken out to draw the line of best fit, merely that the system as a whole has lost two.

Tables of critical values of correlation coefficients can be used without understanding the idea of degrees of freedom, but the idea is an important one which you will often use as you learn more Statistics. In general

$$\text{Degrees of freedom} = \text{Sample size} - \text{Number of restrictions}.$$

Different types of restriction apply in different statistical procedures.

## Interpreting correlation

You need to be on your guard against drawing spurious conclusions from significant correlation coefficients.

### Correlation does not imply causation

Figures for the years 1985–93 show a high correlation between the number of sales of personal computers and those for the sale of microwave ovens. There is of course no direct connection between the two variables. You would be quite wrong to conclude that buying a microwave oven predisposes you to buy yourself a computer as well, or vice versa.

Although there may be a high level of correlation between variables $A$ and $B$ it does not mean that $A \rightarrow B$ or that $A \leftarrow B$. It may well be that a third variable $C$ causes both $A$ and $B$,

or it may be that there is a more complicated set of relationships. In the case of personal computers and microwaves, both are clearly caused by the advance of modern technology.

# Non-linear correlation

A low value of $r$ tells you that there is little or no *linear* correlation. There are other forms of correlation as illustrated in the diagrams of figure 4.7.

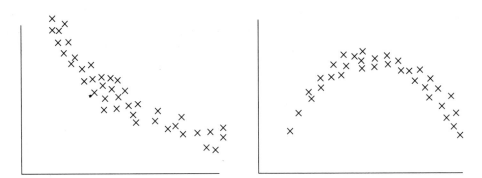

**Figure 4.7 Scatter diagrams showing non-linear correlation**

In these diagrams there is clearly an association between the variables, but not one of linear correlation.

# Extrapolation

A linear relationship established over a particular domain should not be assumed to hold outside this range. For instance, there is strong correlation between the age in years and 100 metre times of female athletes between the ages of 8 and 20 years. To extend the connection, figure 4.8, would suggest that veteran athletes are quicker than athletes who are in their prime, and if they live long enough can even run 100 metres in no time at all!

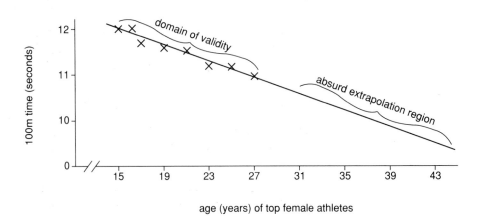

**Figure 4.8**

## Exercise 4C

**4**

Most of the questions in this exercise involve unrealistically small samples. They are meant to help you understand the principles involved in testing for correlation. When you come to do such testing on real data, you would hope to be able to use much larger samples.

**1.** A language teacher wished to test whether there is any correlation between students' ability in their own and a foreign language. Accordingly she collects the marks of 8 students, all native English speakers, in their end of year examinations in English and French.

| Candidate | A | B | C | D | E | F | G | H |
|-----------|---|---|---|---|---|---|---|---|
| **English** | 65 | 34 | 48 | 72 | 58 | 63 | 26 | 80 |
| **French** | 74 | 49 | 45 | 80 | 63 | 72 | 12 | 75 |

(i)   Calculate the product moment correlation coefficient.
(ii)  State the null and alternative hypotheses.
(iii) Using the correlation coefficient as a test statistic, carry out the test at the 5% significance level.

**2.** 'You can't win without scoring goals.' So says the coach of a netball team. Jamila, who believes in solid defensive play, disagrees and sets out to prove that there is no correlation between scoring goals and winning matches. She collects the following data for the goals scored and the points scored by 12 teams in a netball league:

| Goals scored, $x$: | 41 | 50 | 54 | 47 | 47 | 49 | 52 | 61 | 50 | 29 | 47 | 35 |
|--------------------|----|----|----|----|----|----|----|----|----|----|----|----|
| **Points scored, $y$:** | 21 | 20 | 19 | 18 | 16 | 14 | 12 | 11 | 11 | 7 | 5 | 2 |

(i)   Calculate the product moment correlation coefficient.
(ii)  State suitable null and alternative hypotheses, indicating whose position each represents.
(iii) Carry out the hypothesis test and comment on the result.

**3.** A sports reporter believes that those who are good at the high jump are also good at the long jump, and vice versa. He collects data on the best performances of nine athletes, as follows.

| Athlete | A | B | C | D | E | F | G | H | I |
|---------|---|---|---|---|---|---|---|---|---|
| **High jump $x$ metres** | 2.0 | 2.1 | 1.8 | 2.1 | 1.8 | 1.9 | 1.6 | 1.8 | 1.8 |
| **Long jump $y$ metres** | 8.0 | 7.6 | 6.4 | 6.8 | 5.8 | 8.0 | 5.5 | 5.5 | 6.6 |

(i)   Calculate the product moment correlation coefficient.
(ii)  State suitable null and alternative hypotheses.
(iii) Carry out the hypothesis test and comment on the result.

4. It is widely believed that those who are good at chess are good at bridge, and vice versa. A commentator decides to test this theory using the grades of a random sample of 8 people who play both games as data.

| Player | A | B | C | D | E | F | G | H |
|---|---|---|---|---|---|---|---|---|
| Chess grade | 160 | 187 | 129 | 162 | 149 | 151 | 189 | 158 |
| Bridge grade | 75 | 100 | 75 | 85 | 80 | 70 | 95 | 80 |

(i)   Calculate the product moment correlation coefficient.
(ii)  State suitable null and alternative hypotheses.
(iii) Carry out the hypothesis test at the 5% significance level. Do these data support this belief at this significance level?

5. A biologist believes that a particular type of fish develops black spots on its scales in water that is polluted by certain agricultural fertilisers. She catches a number of fish; for each one she counts the number of black spots on its scales and measures the concentration of the pollutant in the water it was swimming in. She uses these data to test for positive linear correlation between the number of spots and the level of pollution.

| Fish | | A | B | C | D | E | F | G | H | I | J |
|---|---|---|---|---|---|---|---|---|---|---|---|
| (Pollutant concentration (parts per million) | | 124 | 59 | 78 | 79 | 150 | 12 | 23 | 45 | 91 | 68 |
| No. of black spots | | 15 | 8 | 7 | 8 | 14 | 0 | 4 | 5 | 8 | 8 |

(i)   Calculate the product moment correlation coefficient.
(ii)  State suitable null and alternative hypotheses.
(iii) Carry out the hypothesis test at the 2% significance level. What can the biologist conclude?

6. Andrew claims that the older you get, the slower is your reaction time. His mother disagrees, saying the two are unrelated. They device that the only way to settle the discussion is to carry out a proper test. A few days later they are having a small party and so ask their twelve guests to take a test that measures their reaction times. The results are as follows:

| Age | Reaction time (secs) | Age | Reaction time (secs.) |
|---|---|---|---|
| 78 | 0.8 | 10 | 0.3 |
| 72 | 0.6 | 19 | 0.3 |
| 60 | 0.7 | 20 | 0.4 |
| 56 | 0.5 | 28 | 0.4 |
| 41 | 0.5 | 30 | 0.3 |
| 39 | 0.4 | 35 | 0.5 |

Carry out the test at the 5% significance level, stating the null and alternative hypotheses. Who won the argument, Andrew or his mother?

7. The teachers at a school have a discussion as to whether girls in general run faster or slower as they get older. They decide to collect data for a random sample of girls the next time the school cross country race is held (which everyone has to take part in). They collect the following data, with the times given in minutes and the ages in years (the conversion from months to decimal parts of a year has already been carried out).

| Age | Time | Age | Time | Age | Time |
|-----|------|-----|------|-----|------|
| 11.6 | 23.1 | 18.2 | 45.0 | 13.9 | 29.1 |
| 15.0 | 24.0 | 15.4 | 23.2 | 18.1 | 21.2 |
| 18.8 | 45.0 | 14.4 | 26.1 | 13.4 | 23.9 |
| 16.0 | 25.2 | 16.1 | 29.4 | 16.2 | 26.0 |
| 12.8 | 26.4 | 14.6 | 28.1 | 17.5 | 23.4 |
| 17.6 | 22.9 | 18.7 | 45.0 | 17.0 | 25.0 |
| 17.4 | 27.1 | 15.4 | 27.0 | 12.5 | 26.3 |
| 13.2 | 25.2 | 11.8 | 25.4 | 12.7 | 24.2 |
| 14.5 | 26.8 | | | | |

(i) State suitable null and alternative hypotheses and decide on an appropriate significance level for the test.

(ii) Calculate the product moment correlation coefficient and state the conclusion from the test.

(iii) Plot the data on a scatter diagram and identify any outliers. Explain how they could have arisen.

(iv) Comment on the validity of the test.

8. The manager of a company wishes to evaluate the success of its training programme. One aspect which interests her is to see if there is any relationship between the amount of training given to employees and the length of time they stay with the company before moving on to jobs elsewhere. She does not want to waste company money training people who will shortly leave. At the same time she believes that the more training employees are given the longer they will stay. She collects data on the average number of days training given per year to 25 employees who have recently left for other jobs, and the length of time they worked for the company.

*Exercise 4c continued*

| Training (days/year) | Work (days) | Training (days/year) | Work (days) | Training (days/year) | Work (days) |
|---|---|---|---|---|---|
| 2.0 | 354 | 3.4 | 760 | 1.2 | 132 |
| 4.0 | 820 | 1.8 | 125 | 4.5 | 1365 |
| 0.1 | 78 | 0.0 | 28 | 1.0 | 52 |
| 5.6 | 1480 | 5.7 | 1360 | 7.8 | 1080 |
| 9.1 | 980 | 7.2 | 1520 | 3.7 | 508 |
| 2.6 | 902 | 7.5 | 1380 | 10.9 | 1281 |
| 0.0 | 134 | 3.0 | 121 | 3.8 | 945 |
| 2.6 | 252 | 2.8 | 457 | 2.9 | 692 |
| 7.2 | 867 | | | | |

(i)   Calculate the product moment correlation coefficient.
(ii)  State suitable null and alternative hypotheses.
(iii) Carry out the hypothesis test at the 5% significance level.
(iv) Plot the data on a scatter diagram.
(v)  What conclusions would you come to if you were the manager?

**9.** Charlotte is a campaigner for temperance, believing that drinking alchohol is an evil habit. Michel, representative of a wine company, presents her with these figures which he claims show that wine drinking is good for marriages.

| Country | Wine consumption (kg/person/year) | Divorce rate (/1000 inhabitants) |
|---|---|---|
| Belgium | 20 | 2.0 |
| Denmark | 20 | 2.7 |
| Germany | 26 | 2.2 |
| Greece | 33 | 0.6 |
| Italy | 63 | 0.4 |
| Portugal | 54 | 0.9 |
| Spain | 41 | 0.6 |
| U.K. | 13 | 2.9 |

(i)   Write Michel's claim in the form of a hypothesis test and carry it out.
(ii)  Charlotte claims that Michel is 'indulging in pseudo-statistics'. What arguments could she use to support this point of view?

**10.** The values of $x$ and $y$ in the table are the marks obtained in an intelligence test and a university examination respectively by 20 medical students. The data are plotted in the scatter diagram.

| $x$ | 98 | 51 | 71 | 57 | 44 | 59 | 75 | 47 | 39 | 58 |
|---|---|---|---|---|---|---|---|---|---|---|
| $y$ | 85 | 40 | 30 | 25 | 50 | 40 | 50 | 35 | 25 | 90 |

| $x$ | 77 | 65 | 58 | 66 | 79 | 72 | 45 | 40 | 49 | 76 |
|---|---|---|---|---|---|---|---|---|---|---|
| $y$ | 65 | 25 | 70 | 45 | 70 | 50 | 40 | 20 | 30 | 60 |

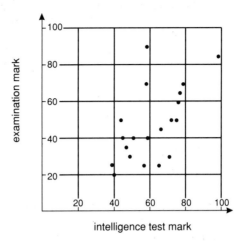

Given that $\Sigma x = 1226$, $\Sigma y = 945$, $\Sigma x^2 = 79\,732$, $\Sigma y^2 = 52\,575$ and $\Sigma xy = 61\,495$, calculate the product moment correlation coefficient $r$ to 2 decimal places.

Referring to the evidence provided by the diagram and the value of $r$, comment briefly on the correlation between the two sets of marks.

Now eliminate from consideration those 10 students whose values of $x$ are less than 50 or more than 75. Calculate the new value of $r$ for the marks of the remaining students. What does the comparison with the earlier value of $r$ seem to indicate?                [MEI]

**11.** The following data refer to the price (in £/kg) of mammoth meat in London for the four quarters of the years 1987–92.

| Year | Quarter | Price | Year | Quarter | Price |
|------|---------|-------|------|---------|-------|
| 1992 | 4 | £5.00 | 1989 | 4 | £4.70 |
|      | 3 | £6.00 |      | 3 | £5.70 |
|      | 2 | £6.90 |      | 2 | £6.80 |
|      | 1 | £6.10 |      | 1 | £5.80 |
| 1991 | 4 | £4.90 | 1988 | 4 | £4.50 |
|      | 3 | £5.80 |      | 3 | £5.60 |
|      | 2 | £6.90 |      | 2 | £6.50 |
|      | 1 | £5.80 |      | 1 | £5.40 |
| 1990 | 4 | £4.80 | 1987 | 4 | £4.40 |
|      | 3 | £5.80 |      | 3 | £5.50 |
|      | 2 | £6.80 |      | 2 | £6.30 |
|      | 1 | £5.90 |      | 1 | £5.20 |

*Exercise 4c continued*

(i) Form these data into a bivariate set by taking each price with the one before it: (5.00,6.00), (6.00,6.90), (6.90,6.10), ..., (6.30,5.20). Work out the correlation coefficient for this set and call it $r_1$.

(ii) Form another bivariate set by doing the same thing but with data from two quarters apart: (5.00,6.90), (6.00,6.10), ..., (5.50,5.20). Work out the correlation coefficient for this set and call it $r_2$.

(iii) Repeat this procedure for the sets of bivariate data 3 quarters apart and find $r_3$.

(iv) Repeat this procedure for the sets of bivariate data 4 quarters apart and find $r_4$.

(v) What do you notice? This technique of correlating time-dependent data with itself is called *autocorrelation* and is used to detect cyclic or seasonal variations in the data. What do the different values of $r$ tell you about these data.

**12.** Examine this newspaper report critically.

# British publishers hold the key to continental exchange rates

Startling new evidence has come to light of the effect of the flourishing British publishing industry on the economies of France and Germany. The figures for UK Book Title Production and for the real Franc/Mark exchange rate (adjusted for differentials in producer prices) are given below for the period 1980–1990.

| Year | 1980 | 1981 | 1982 | 1983 | 1984 | 1985 | 1986 | 1987 | 1988 |
|------|------|------|------|------|------|------|------|------|------|
| Books | 48158 | 43083 | 48307 | 51071 | 51555 | 52994 | 52496 | 54746 | 56514 |
| Exch. Rate | 96 | 93 | 95 | 100 | 101 | 97 | 99 | 105 | 105 |

| Year | 1989 | 1990 |
|------|------|------|
| Books | 61361 | 63756 |
| Exch. Rate | 100 | 104 |

The correlation between the two is significant even at the 1% Level, showing beyond all reasonable doubt that our publishers are steadily undermining the German currency.

(Note. The article is fictitious but the figures are real.)

THE AVONFORD STAR

# Punch-up at village fete

Pandemonium broke out at the Normanton village fete last Saturday when the adjudication for the Tomato of the Year competition was announced. The two judges completely failed to agree in their rankings and so a compromise winner was chosen to the fury of everybody (except the winner).

Following the announcement there was a moment of stunned silence, followed by shouts of "Rubbish", "It's a fix", "Go home" and further abuse. Then the tomatoes started to fly and before long fighting broke out.

By the time the police arrived on the scene ten people were injured, including last year's winner Bert Wallis who lost three teeth in the scrap. Both judges had escaped unhurt.

Angry Bert Wallis, nursing a badly bruised jaw, said "The competition was a nonsense. The judges were useless. Their failure to agree shows they did not know what they were looking for". But fete organiser Margaret Bramble said this was completely untrue. "The competition was absolutely fair; both judges know a good tomato when they see one," she claimed.

The judgement that caused all the trouble was as follows:

| Tomato | A | B | C | D | E | F | G | H |
|--------|---|---|---|---|---|---|---|---|
| Judge 1 | 1 | 8 | 4 | 6 | 2 | 5 | 7 | 3 |
| Judge 2 | 7 | 2 | 3 | 4 | 6 | 8 | 1 | 5 |
| Total | 8 | 10 | 7 | 10 | 8 | 13 | 8 | 8 |
|  |  |  | Winner |  |  |  |  |  |

You will see that both judges ranked the eight entrants, 1st, 2nd 3rd, ..., 8th. The winner, C, was placed 4th by one judge and 3rd by the other. Their rankings look different so perhaps they were using different criteria on which to assess them. How can you use these data to decide whether that was or was not the case?

One way would be to calculate a correlation coefficient and use it to carry out a hypothesis test:

$$H_0: \quad \rho = 0 \quad \text{There is no correlation}$$

$$H_1: \quad \rho > 0 \quad \text{There is positive correlation}$$

The null hypothesis, $H_0$, represents something like Bert Wallis's view, the alternative hypothesis that of Margaret Bramble.

However the data you have are of a different type from any that you have used before for calculating correlation coefficients. In the point (1,7), corresponding to tomato A, the numbers 1 and 7 are *ranks* and not scores, like marks in an examination or measurements. It is however possible to calculate a *rank correlation coefficient*, and in the same way as before.

| Tomato | Judge 1 $x_i$ | Judge 2 $y_i$ | $x_i^2$ | $y_i^2$ | $x_i y_i$ |
|---|---|---|---|---|---|
| A | 1 | 7 | 1 | 49 | 7 |
| B | 8 | 2 | 64 | 4 | 16 |
| C | 4 | 3 | 16 | 9 | 12 |
| D | 6 | 4 | 36 | 16 | 24 |
| E | 2 | 6 | 4 | 36 | 12 |
| F | 5 | 8 | 25 | 64 | 40 |
| G | 7 | 1 | 49 | 1 | 7 |
| H | 3 | 5 | 9 | 25 | 15 |
| Total $\Sigma$ | 36 | 36 | 204 | 204 | 133 |

$$\bar{x} = \frac{\Sigma x_i}{n} = \frac{36}{8} = 4.5 \qquad \text{Similarly } \bar{y} = \frac{36}{8} = 4.5$$

$$s_x = \sqrt{\left(\frac{1}{n}\Sigma x_i^2 - \bar{x}^2\right)} = \sqrt{\left(\frac{204}{8} - 4.5^2\right)} = 2.29$$

Similarly $s_y = 2.29$

$$s_{xy} = \frac{1}{n}\Sigma x_i y_i - \bar{x}\bar{y} = \frac{133}{8} - 4.5 \times 4.5 = -3.625 \; Covariance$$

$$r = \frac{s_{xy}}{s_x s_y} = \frac{-3.625}{2.29 \times 2.29} = -0.69 \; \begin{array}{l}Correlation \\ coefficient\end{array}$$

Since the correlation coefficient is negative there can be no possibility of accepting $H_1$, therefore you accept $H_0$. There is no evidence of agreement between the judges. Remember this was a 1-tail test for positive correlation.

## Spearman's Coefficient of Rank Correlation

The calculation in the previous example is usually carried out by a quite different, but equivalent, procedure and the resulting correlation coefficient is called *Spearman's coefficient of rank correlation* and denoted by $r_s$.

The procedure is summarised by the formula

$$r_s = 1 - \frac{6\Sigma d_i^2}{n(n^2 - 1)}$$

where $d_i$ is the difference in the ranks given to the $i^{th}$ item.

The calculation is then as follows:

| Tomato | Judge 1 $x_i$ | Judge 2 $y_i$ | $d_i = x_i - y_i$ | $d_i^2$ |
|---|---|---|---|---|
| A | 1 | 7 | −6 | 36 |
| B | 8 | 2 | 6 | 36 |
| C | 4 | 3 | 1 | 1 |
| D | 6 | 4 | 2 | 4 |
| E | 2 | 6 | −4 | 16 |
| F | 5 | 8 | −3 | 9 |
| G | 7 | 1 | 6 | 36 |
| H | 3 | 5 | −2 | 4 |
| | | | Total $\Sigma d_i^2$ | 142 |

$$r_s = 1 - \frac{6\Sigma d_i^2}{n(n^2 - 1)} = 1 - \frac{6 \times 142}{8(8^2 - 1)}$$

$$= 1 - 1.690$$

$$= -0.690$$

You will see that this is the same answer as before, but the working is much shorter. It is not difficult to prove that the two methods are equivalent, and this is done on page 130 of the Appendix.

Critical values for Spearman's rank correlation coefficient are however different from those for Pearson's product moment correlation coefficient, so you must always be careful to use the appropriate tables.

The calculation is often carried out with the data across the page rather than in columns and this is shown in the next example.

---

**EXAMPLE**

During their course two trainee tennis coaches, Rachael and Leroy, were shown videos of 7 people, A, B, C, ..., G, doing a top spin service, and were asked to rank them in order according to the quality of their style. They placed them as follows:

| Rank order | 1 | 2 | 3 | 4 | 5 | 6 | 7 |
|---|---|---|---|---|---|---|---|
| Rachael | B | G | F | D | A | C | E |
| Leroy | F | B | D | E | G | A | C |

(i)  Find Spearman's coefficient of rank correlation.

(ii) Use it to test whether there is evidence at the 5% level of positive correlation between their judgements.

*Solution*

(i) **The rankings are:**

|  | A | B | C | D | E | F | G |  |
|---|---|---|---|---|---|---|---|---|
| **Rachael** | 5 | 1 | 6 | 4 | 7 | 3 | 2 |  |
| **Leroy** | 6 | 2 | 7 | 3 | 4 | 1 | 5 |  |
| $d_i$ |  | −1 | −1 | −1 | 1 | 3 | 2 | −3 | $n = 7$ |
| $d_i^2$ |  | 1 | 1 | 1 | 1 | 9 | 4 | 9 | $\Sigma d_i^2 = 26$ |

$$r_s = 1 - \frac{6\Sigma d_i^2}{n(n^2 - 1)} = 1 - \frac{6 \times 26}{7(7^2 - 1)}$$

$$= 0.54 \text{ to 2 decimal places}$$

(ii) $H_0$: $\rho = 0$ There is no correlation between their rankings

$H_1$: $\rho > 0$ There is positive correlation between their rankings

Significance level 5% 1-tail test

From tables the critical value of $r_s$ for a 1-tail test at this significance level, for $n = 7$ is 0.7143.

$$0.54 < 0.7143$$

So $H_0$ is accepted.

**Critical values for Spearman's rank correlation coefficient, $r_s$**

| | 5% | 2½% | 1% | ½% | 1–Tail Test |
|---|---|---|---|---|---|
| | 10% | 5% | 2% | 1% | 2–Tail Test |
| $n$ | | | | | |
| 1 | – | – | – | –· | |
| 2 | – | – | – | – | |
| 3 | ·– | – | – | – | |
| 4 | 1.0000 | – | – | – | |
| 5 | 0.9000 | 1.0000 | 1.0000 | – | |
| 6 | 0.8286 | 0.8857 | 0.9429 | 1.0000 | |
| 7 | 0.7143 | 0.7857 | 0.8929 | 0.9286 | |
| 8 | 0.6429 | 0.7381 | 0.8333 | 0.8810 | |
| 9 | 0.6000 | 0.7000 | 0.7833 | 0.8333 | |
| 10 | 0.5636 | 0.6485 | 0.7455 | 0.7939 | |

*Extract from table of values for Spearman's rank correlation coefficient, $r_s$*

There is insufficient evidence to claim positive correlation between their rankings.

**HISTORICAL NOTE**

*Charles Spearman was born in London in 1863. After serving 14 years in the army as a cavalry officer, he went to Leipzig to study psychology. On completing his doctorate there he became a lecturer, and soon afterwards a professor, at University College, London. He pioneered the application of statistical techniques within pyschology and developed the technique known as factor analysis in order to analyse different aspects of human ability. He died in 1945.*

## Tied Ranks

If several items are ranked equally you give them the mean of the ranks they would have had if they had been slightly different from each other.

For example A, B, …, J are ranked

| 1 | 2= | 2= | 4 | 5 | 6= | 6= | 6= | 0 | 10 |
|---|----|----|---|---|----|----|----|---|----|
| C | G  | J  | A | D | B  | I  | F  | E | H  |

G and J are both 2= and so are given the rank $\frac{1}{2}(2+3) = 2.5$.

B, I and F are all 6= and so are given the rank $\frac{1}{3} \times (6+7+8) = 7$.

## When to use rank correlation

Sometimes your data will be available in two forms, as values of variables or in rank order. If you have the choice you will usually work out the correlation coefficient from the variable values rather than the ranks.

If may well be the case however that only ranked data is available to you and in that case you would have no choice but to use it. It may also be that while you could collect variable values as well, it would not be worth the time, trouble or expense to do so.

Pearson's product moment correlation coefficient is a measure of *linear* correlation and so is not appropriate for non-linear data like those illustrated in the scatter diagram, figure 4.9. You may however use rank correlation to investigate whether one variable generally increases (or decreases) as the other increases.

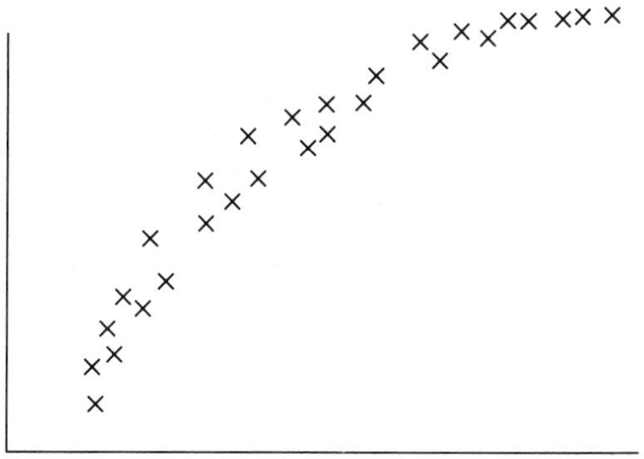

**Figure 4.9 Non-linear data with a high degree of rank correlation**

You should however always look at the sense of your data before deciding which is the more appropriate form of correlation to use.

**NOTE** *Spearman's rank correlation coefficient provides one among many statistical tests that can be carried out on ranks rather than variable values. Such tests are examples of non-parametric tests. A non-parametric test is a test on some aspect of a distribution which is not specified by its defining parameters.*

## Exercise 4D

**1.** Two judges in a baby competition rank the 8 babies as follows:

| Baby | A | B | C | D | E | F | G | H |
|------|---|---|---|---|---|---|---|---|
| Judge 1 | 2 | 6 | 1 | 5 | 7 | 4 | 3 | 8 |
| Judge 2 | 3 | 1 | 5 | 2 | 7 | 8 | 4 | 6 |

Calculate Spearman's rank correlation coefficient between the judges and comment on your result. [SUJB]

**2.** The order of merit of 10 individuals at the start and finish of a training course were:

| Individual | A | B | C | D | E | F | G | H | I | J |
|------------|---|---|---|---|---|---|---|---|---|---|
| Order at start | 1 | 2 | 3 | 4 | 5 | 6 | 7 | 8 | 9 | 10 |
| Order at finish | 5 | 3 | 1 | 9 | 2 | 6 | 4 | 7 | 10 | 8 |

Find Spearman's coefficient of rank correlation between the two orders.

**3.** Two adjudicators at a music competition awarded marks to 10 pianists as follows:

| Pianist | A | B | C | D | E | F | G | H | I | J |
|---------|---|---|---|---|---|---|---|---|---|---|
| Adjudicator 1 | 78 | 66 | 73 | 73 | 84 | 66 | 89 | 84 | 67 | 77 |
| Adjudicator 2 | 81 | 68 | 81 | 75 | 80 | 67 | 85 | 83 | 66 | 78 |

Calculate a coefficient of rank correlation for these data. [SUJB]

**4.** In a skating competition two judges place the contestants, in descending order of merit, as follows:

| Judge 1 | C | E | D | F | A | G | I | J | B | H |
|---------|---|---|---|---|---|---|---|---|---|---|
| Judge 2 | F | G | D | A | I | C | H | E | J | B |

(i) Find Spearman's coefficient of rank correlation between the two orders.

(ii) Stating suitable null and alternative hypotheses, test at the 5% significance level whether the judges were in broad agreement with each other.

**5.** An education psychologist obtained scores by 9 university entrants in 3 tests (A, B and C). The scores in tests A and B were as follows:

| Entrant | 1 | 2 | 3 | 4 | 5 | 6 | 7 | 8 | 9 |
|---------|---|---|---|---|---|---|---|---|---|
| A score | 8 | 3 | 9 | 10 | 4 | 9 | 6 | 4 | 5 |
| B score | 7 | 8 | 5 | 9 | 10 | 6 | 3 | 4 | 7 |

Calculate a coefficient of rank correlation between the two sets of scores. The coefficient obtained between the A and C scores was 0.71 and that between the B and C scores was 0.62. What advice would you give the psychologist if he wished to use less than three tests?          [C]

6. In a driving competition there were eight contestants and three judges who placed them in rank order as shown in the table below:

| Competitor | A | B | C | D | E | F | G | H |
|---|---|---|---|---|---|---|---|---|
| Judge X | 2 | 5 | 6 | 1 | 8 | 4 | 7 | 3 |
| Judge Y | 1 | 6 | 8 | 3 | 7 | 2 | 4 | 5 |
| Judge Z | 2= | 2= | 6= | 4 | 6= | 1 | 6= | 5 |

(i)  Which two judges agreed the most?
(ii) Stating suitable null and alternative hypotheses, carry out a hypothesis test on the level of agreement these two judges.

7. In a marrow competition there were seven entrants and two judges who made the following decisions:

| Marrow | A | B | C | D | E | F | G |
|---|---|---|---|---|---|---|---|
| Judge X | 6 | 1 | 7 | 4 | 2 | 5 | 3 |
| Judge Y | 1 | 6 | 2 | 5 | 7 | 4 | 3 |

Calculate Spearman's coefficient of rank correlation between the judges. What can you conclude from your answer?

8. A coach wanted to test his theory that, although athletes have specialisms, it is still true that those who run fast at one distance are also likely to run fast at another distance. He selected six athletes at random to take part in a test and invited them to compete over 100m and over 1500m.

The times and places of the six athletes were as follows:

| Athlete | 100m time | 100m rank | 1500m time | 1500m rank |
|---|---|---|---|---|
| Allotey | 9.8 s | 1 | 3 m 42 s | 1 |
| Chell | 10.9 s | 6 | 4 m 11 s | 2 |
| Giles | 10.4 s | 2 | 4 m 19 s | 6 |
| Mason | 10.5 s | 3 | 4 m 18 s | 5 |
| O'Hara | 10.7 s | 5 | 4 m 12 s | 3 |
| Stuart | 10.6 s | 4 | 4 m 16 s | 4 |

(i)  Calculate the Pearson product moment and Spearman's rank correlation coefficient for these data.
(ii) State suitable null and alternative hypotheses and carry out hypothesis tests on these data.

(iii) State which you consider to be the more appropriate correlation coefficient in this situation, giving your reasons.

9. In a random sample of 8 areas, residents were asked to express their approval or disapproval of the services provided by the local authority. A score of 0 represented complete dissatisfaction, and 10 represented complete satisfaction. The table below shows the mean score for each local authority together with the authority's level of community charge.

| Authority | A | B | C | D | E | F | G | H |
|---|---|---|---|---|---|---|---|---|
| Community Charge (£) | 485 | 490 | 378 | 451 | 384 | 352 | 420 | 212 |
| Approval rating | 3.0 | 4.4 | 5.0 | 4.6 | 4.1 | 5.5 | 5.8 | 6.1 |

Calculate Spearman's rank correlation coefficient for the data.

Carry out a significance test at the 5% level using the value of the correlation coefficient which you have calculated. State carefully the null and alternative hypotheses under test and the conclusion to be drawn.

[MEI]

10. At the end of a word-processing course the trainees are given a document to type. They are assessed on the time taken and on the quality of their work. For a random sample of 12 trainees the following results were obtained.

| Trainee | A | B | C | D | E | F | G | H | I | J | K | L |
|---|---|---|---|---|---|---|---|---|---|---|---|---|
| Quality (%) | 97 | 96 | 94 | 91 | 90 | 87 | 86 | 83 | 82 | 80 | 77 | 71 |
| Time (seconds) | 210 | 230 | 198 | 204 | 213 | 206 | 200 | 186 | 192 | 202 | 191 | 199 |

(i) Calculate Spearman's coefficient of rank correlation for the data. Explain what the sign of your correlation coefficient indicates about the data.

(ii) Carry out a test, at the 5% level of significance, of whether or not there is any correlation between time taken and quality of work for trainees who have attended this word-processing course. State clearly the null and alternative hypotheses under test and the conclusion reached.

[MEI]

11. A school holds an election for parent governors. Candidates are invited to write brief autobiographies and these are sent out at the same time as the voting papers.

After the election, one of the candidates, Mr. Smith, says that the more words you wrote about yourself the more votes you got. He sets out to 'prove this statistically' by calculating the product moment correlation between the numbers of words and the numbers of votes.

*Exercise 4d continued*

| Candidate | A | B | C | D | E | F | G |
|---|---|---|---|---|---|---|---|
| **Number of words:** | 70 | 101 | 106 | 232 | 150 | 102 | 98 |
| **Number of votes:** | 99 | 108 | 97 | 144 | 94 | 54 | 87 |

(i) Calculate the product moment correlation coefficient.

Mr. Smith claims that this proves his point at the 5% significance level.

(ii) State his null and alternative hypotheses and show how he came to his conclusion.

(iii) Calculate Spearman's rank correlation coefficient for these data.

(iv) Explain the difference in the two correlation coefficients and criticise the procedure Mr. Smith used for coming to his conclusion.

**12.** These data, referring to the ordering of perceived risk for 25 activities and technologies and actual fatality estimates, were obtained in a study in the United States. Use these data to test at the 5% significance level for positive correlation between:

(i) The League of Women Voters and College students;

(ii) Experts and Actual fatality estimates;

(iii) College students and Experts.

Comment on your results and identify any outliers in the three sets of bivariate data you have just used.

| | League of Women Voters | College students | Experts | Actual fatalities (estimates) |
|---|---|---|---|---|
| **Nuclear power** | 1 | 1 | 18 | 16 |
| **Motor vehicles** | 2 | 4 | 1 | 3 |
| **Handguns** | 3 | 2 | 4 | 4 |
| **Smoking** | 4 | 3 | 2 | 1 |
| **Motorcycles** | 5 | 5 | 6 | 6 |
| **Alcoholic beverages** | 6 | 6 | 3 | 2 |
| **General (private) aviation** | 7 | 12 | 11 | 11 |
| **Police work** | 8 | 7 | 15 | 18 |
| **Surgery** | 9 | 10 | 5 | 8 |
| **Fire fighting** | 10 | 9 | 16 | 17 |
| **Large construction** | 11 | 11 | 12 | 12 |
| **Hunting** | 12 | 15 | 20 | 14 |
| **Mountain climbing** | 13 | 17 | 24 | 21 |
| **Bicycles** | 14 | 19 | 13 | 13 |
| **Commercial aviation** | 15 | 13 | 14 | 20 |
| **Electric power (non-nuclear)** | 16 | 16 | 8 | 5 |
| **Swimming** | 17 | 25 | 9 | 7 |
| **Contraceptives** | 18 | 8 | 10 | 19 |

| | League of Women Voters | College students | Experts | Actual fatalities (estimates) |
|---|---|---|---|---|
| **Skiing** | 19 | 20 | 25 | 24 |
| **X-rays** | 20 | 14 | 7 | 9 |
| **High school & college football** | 21 | 21 | 22 | 23 |
| **Railroads** | 22 | 18 | 17 | 10 |
| **Power mowers** | 23 | 23 | 23 | 22 |
| **Home appliances** | 24 | 22 | 19 | 15 |
| **Vaccinations** | 25 | 24 | 21 | 25 |

Source: Schwing and Albers, *Societal Risk Assessment*, Plenum

13. To test the belief that milder winters are followed by warmer summers, meteorological records are obtained for a random sample of 10 years. For each year the mean temperatures are found for January and July. The data, in degrees Celsius, are given below.

| **January** | 8.3 | 7.1 | 9.0 | 1.8 | 3.5 | 4.7 | 5.8 | 6.0 | 2.7 | 2.1 |
|---|---|---|---|---|---|---|---|---|---|---|
| **July** | 16.2 | 13.1 | 16.7 | 11.2 | 14.9 | 15.1 | 17.7 | 17.3 | 12.3 | 13.4 |

(i) Rank the data and calculate Spearman's rank correlation coefficient.
(ii) Test, at the 2.5% level of significance, the belief that milder winters are followed by warmer summers. State clearly the null and alternative hypotheses under test.
(iii) Would it be more appropriate, less appropriate or equally appropriate to use the product moment correlation coefficient to analyse these data? Briefly explain why. [MEI]

# The Least Squares Regression Line

At the start of this chapter you saw how to draw a line of best fit through a set of points on a scatter diagram by eye. You did this by drawing it through the mean point, leaving roughly the same number of points above and below it. This is obviously a very imprecise method.

Before you do any calculations you first need to look carefully at the two variables that give rise to your data. It is normal practice to plot the *dependent variable* on the vertical axis and the *independent variable* on the horizontal axis. In the example which follows, the independent variable is the time at which measurements are made. (Notice that this is a non-random variable). The procedure leads to the equation of the *regression line*, the line of best fit in these circumstances.

Look at the scatter diagram (figure 4.10) showing the $n$ points A $(x_1, y_1)$, B $(x_2, y_2)$, ..., N $(x_n, y_n)$. On it is marked a possible line of best fit $l$. If the $l$ passed through all the points there would be no problem since there would be perfect linear correlation. it does not of course pass though all the points and you would be very surprised if such a line did in any real situation.

By how much is it missing the points? The answer to that question is shown by the five vertical lines from the points to the line. Their lengths $\varepsilon_1$, $\varepsilon_2$ ...,$\varepsilon_n$ are called the *residuals* and represent the variation which is not explained by the line $l$. The *least squares regression line* is the line which produces the least possible value of the sum of the squares of the residuals, $\varepsilon_1^2 + \varepsilon_2^2 + \ldots + \varepsilon_n^2$.

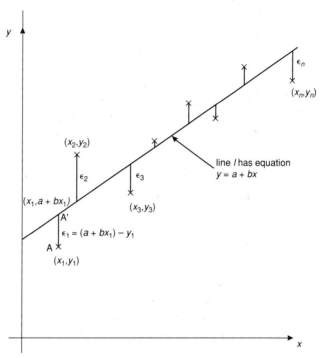

**Figure 4.10 Bivariate data plotted on a scatter diagram with the regression line, $l$, $y=a+bx$, and the residuals $\varepsilon_1$, $\varepsilon_2$, ..., $\varepsilon_n$**

If the equation of the line $l$ is

$$y = a + bx,$$

then it is easy to see that the point A′ on the diagram, directly above A, has coordinates

$$(x_1, a+bx_1)$$

and so the corresponding residual, $\varepsilon_1$, is given by

$$\varepsilon_1 = (a+bx_1) - y_1.$$

Similarly for $\varepsilon_2$, $\varepsilon_3$, ..., $\varepsilon_n$.

The problem is to find the values of the constants $a$ and $b$ in the equation of the line $l$ which make $\varepsilon_1^2 + \varepsilon_2^2 + \ldots + \varepsilon_n^2$ a minimum for any particular set of data, that is to minimise

$$[(a+bx_1)-y_1]^2 + [(a+bx_2)-y_2]^2 + \ldots + [(a+bx_n)-y_n]^2$$

The mathematics involved in doing this is not particularly difficult and is given on page 131 of the Appendix. The resulting equation of the regression line is usually written in the form

$$y - \bar{y} = \frac{S_{xy}}{S_x^2}(x - \bar{x})$$

**NOTES**

1. *In the preceding work you will see that only variation in the y values has been considered. The reason for this is that the x values represent a non-random variable. That is why the residuals are vertical and not in any other direction. Thus $y_1$, $y_2$ ... are values of a random variable Y given by $Y = a + bx + \varepsilon$ where $\varepsilon$ is the residual variation, the variation that is not explained by the regression line.*

2. *The goodness of fit of a regression line may be judged by eye by looking at a scatter diagram. An informal measure which is often used is the Coefficient of Determination, $r^2$, which measures the proportion of the total variation in the dependent variable, Y, which is accounted for by the regression line. There is no standard hypothesis test based on the Coefficient of Determination.*

3. *This form of the regression line is often called the 'y on x regression line'. If for some reason you had y as your independent variable, you would use the 'x on y' form obtained by interchanging x and y in the equation.*

4. *Although the derivation given on page 131 is only true for a random variable on a non-random variable, it happens that for quite different reasons the same form of the regression line applies if both variables are random and Normally distributed. Since this is a common situation, this form of the regression line may be used more widely than might at first have seemed to be the case.*

**EXAMPLE**

A patient is given a drip feed containing a particular chemical and its concentration in his blood is measured, in suitable units, at one hour intervals for the next 6 hours. The doctors believe the figures to be subject to random errors, arising both from the sampling procedure and the subsequent chemical analysis, but that a linear model is appropriate.

| Time, $x$ (hours): | 0 | 1 | 2 | 3 | 4 | 5 | 6 |
|---|---|---|---|---|---|---|---|
| Concentration, $y$: | 2.4 | 4.3 | 5.0 | 6.9 | 9.1 | 11.4 | 13.5 |

(i) Find the equation of the regression line of $y$ upon $x$.

(ii) Estimate the concentration of the chemical in the patient's blood $3\frac{1}{2}$ hours after treatment started.

*Solution*

(i)

| $x_i$ | $y_i$ | $x_i^2$ | $y_i^2$ | $x_iy_i$ | |
|-------|-------|---------|---------|----------|---|
| 0 | 2.4 | 0 | 5.76 | 0 | |
| 1 | 4.3 | 1 | 18.49 | 4.3 | |
| 2 | 5.0 | 4 | 25.00 | 10.0 | |
| 3 | 6.9 | 9 | 47.61 | 20.7 | $n = 7$ |
| 4 | 9.1 | 16 | 82.81 | 36.4 | |
| 5 | 11.4 | 25 | 129.96 | 57.0 | |
| 6 | 13.5 | 36 | 182.25 | 81.0 | |
| $\Sigma$  21 | 52.6 | 91 | 491.88 | 209.4 | |

$$\bar{x} = \frac{\Sigma x_i}{n} = \frac{21}{7} = 3 \quad \bar{y} = \frac{\Sigma y_i}{n} = \frac{52.6}{7} = 7.514$$

$$s_x^2 = \frac{1}{n}\Sigma x_i^2 - \bar{x}^2 = \frac{91}{7} - 3^2 = 4.0$$

$$s_{xy} = \frac{1}{n}\Sigma x_i y_i - \bar{x}\bar{y} = \frac{209.4}{7} - 3 \times 7.514 = 7.372 \text{ Covariance}$$

The regression line is given by

$$y - \bar{y} = \frac{s_{xy}}{s_x^2}(x - \bar{x})$$

$$y - 7.514 = \frac{7.372}{4.0}(x - 3)$$

or $\quad y = 1.843x + 1.985$

(ii)  when $x = 3.5$, $\quad y = 1.843 \times 3.5 + 1.985$

$$= 8.4 \text{ (to 1 decimal place)}$$

At time $3\frac{1}{2}$ hours the concentration is estimated to be 8.4 units.

**N O T E** *The concentration of 8.4 lies between the measured values of 6.9 at time 3 hours and 9.1 at time 4 hours and so seems quite reasonable.*

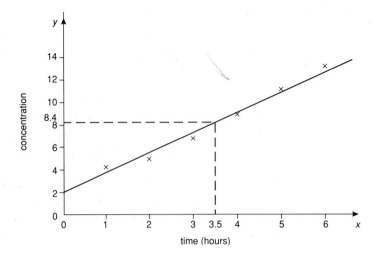

## Exercise 4E

1. For the following bivariate data obtain the equation of the least squares regression line of $y$ on $x$. Estimate the value of $y$ when $x = 12$.

| $x$ | 5 | 10 | 15 | 20 | 25 |
|-----|----|----|----|----|----|
| $y$ | 30 | 28 | 27 | 27 | 21 |

2. Calculate the equation of the regression line of $y$ on $x$ for the following distribution and use it to estimate the value of $y$ when $x = 42$.

| $x$ | 25 | 30 | 35 | 40 | 45 | 50 |
|-----|----|----|----|----|----|----|
| $y$ | 78 | 70 | 65 | 58 | 48 | 42 |

3. The 1970 and 1990 catalogue prices in pence of five British postage stamps are as follows:

| 1970 price, $x$: | 10 | 20 | 30 | 40 | 50 |
|-----|----|----|----|----|----|
| 1990 price, $y$: | 100 | 215 | 280 | 360 | 450 |

(i) Plot these data on a scatter diagram.
(ii) Calculate the equation of the regression line and draw it accurately on your scatter diagram.
(iii) Another stamp was valued at £50 in 1970 and at £120 in 1990. Comment.

4. In an investigation of the genus *Tamarix* ( a shrub able to withstand drought), research workers in Tunisia measured the average vigour $y$ (defined as the average width in cm of the last two annual rings) and stem density $x$ (defined as the number of stems per m²) at 10 sites with the following results:

| x | 4 | 5 | 6 | 9 | 14 | 15 | 15 | 19 | 21 | 22 |
|---|---|---|---|---|----|----|----|----|----|----|
| y | 0.75 | 1.20 | 0.55 | 0.60 | 0.65 | 0.55 | 0 | 0.35 | 0.45 | 0.40 |

(i) Draw a scatter diagram for these data.

(ii) Given that $\Sigma x = 130$, $\Sigma x^2 = 2090$, $\Sigma y = 5.5$ and $\Sigma xy = 59.95$, find the regression line of $y$ on $x$, and plot it on your diagram.

(iii) Use your line to estimate the average vigour for a stem density of 17 stems per m².

(iv) Give one reasons why it would be invalid to use your regression line to estimate average vigour for stem densities substantially greater than 22 stems per m², and explain how your regression line confirms that your reason is correct. [MEI]

5. Each of a group of 12 apple trees is given the same spraying treatment against codling moth larvae. After the crop is gathered, the apples are examined for grubs with the following results.

| Tree number | 1 | 2 | 3 | 4 | 5 | 6 | 7 | 8 | 9 | 10 | 11 | 12 |
|---|---|---|---|---|---|---|---|---|---|---|---|---|
| Size of crop $x$ (hundreds of apples) | 8 | 6 | 11 | 22 | 14 | 17 | 18 | 24 | 19 | 23 | 26 | 40 |
| Percentage $y$ of apples with grubs | 59 | 58 | 56 | 53 | 50 | 45 | 43 | 42 | 39 | 38 | 30 | 27 |

Plot the data on a scatter diagram.

Given that $\Sigma x = 228$, $\Sigma x^2 = 5256$, $\Sigma y = 540$ and $\Sigma xy = 9324$, calculate the line of regression of $y$ on $x$, and plot the line on your diagram. Estimate to the nearest whole number the expected percentage of apples with grubs for a tree carrying 2000 apples. (You may assume that these procedures are justified because $x$ and $y$ are jointly distributed in a suitable way.)

A second group of 12 apple trees is given a different treatment, and the results are given below.

| x | 15 | 15 | 12 | 26 | 18 | 12 | 8 | 38 | 26 | 19 | 29 | 22 |
|---|----|----|----|----|----|----|---|----|----|----|----|----|
| y | 52 | 46 | 38 | 37 | 37 | 37 | 34 | 25 | 22 | 22 | 20 | 14 |

Plot the data from the second group on the same scatter diagram using a different symbol from that used before. *Without further calculation,* summarize the differences between the results of the two treatments. [MEI]

6. In a certain state in the U.S.A. radioactive wastes from a plutonium plant leaked into a river flowing into the ocean. For each of 9 counties bounded by the river or the ocean (or both) an index $x$ of exposure to

*Exercise 4e continued*

radioactivity was calculated. The table shows the values of $x$ for each of these counties together with the corresponding values of $y$, the mortality from cancer per 100 000 person-years, for the period from 1959–1964 inclusive.

| County | Index of exposure: $x$ | Cancer mortality: $y$ |
|--------|----------------------|----------------------|
| 1 | 8.34 | 210.3 |
| 2 | 6.41 | 177.9 |
| 3 | 3.41 | 129.9 |
| 4 | 3.83 | 162.3 |
| 5 | 2.57 | 130.1 |
| 6 | 11.64 | 207.5 |
| 7 | 1.25 | 113.5 |
| 8 | 2.49 | 147.1 |
| 9 | 1.62 | 137.5 |

Calculate the means of $x$ and $y$.

Plot the data on a scatter diagram.

By performing an appropriate calculation, which should be justified, estimate the cancer mortality per 100 000 person-years that would have been observed in a county with an index of exposure of 7.00.     [MEI]

7. The following table shows the weight losses ($y$ kg) of 15 women who had followed a weight-reducing course for various numbers of weeks ($x$).

| Woman | $x$ | $y$ | Woman | $x$ | $y$ | Woman | $x$ | $y$ |
|-------|-----|-----|-------|-----|-----|-------|-----|-----|
| 1 | 6 | 3.3 | 6 | 8 | 5.9 | 11 | 10 | 7.8 |
| 2 | 7 | 4.2 | 7 | 8 | 5.6 | 12 | 11 | 8.2 |
| 3 | 8 | 5.3 | 8 | 9 | 6.6 | 13 | 12 | 9.5 |
| 4 | 8 | 4.8 | 9 | 10 | 6.8 | 14 | 12 | 9.2 |
| 5 | 8 | 6.2 | 10 | 10 | 7.4 | 15 | 14 | 11.9 |

You are given that

$$\Sigma x = 141, \quad \Sigma y = 102.7, \quad \Sigma x^2 = 1391, \quad \Sigma y^2 = 773.41, \quad \Sigma xy = 1032.2.$$

(i) Calculate the correlation coefficient and test whether it is significantly different from zero.
(ii) Plot the data as a scatter diagram on graph paper.
(iii) Calculate the equation of the sample least squares line expressing weight loss ($y$) as a linear function of the number of weeks ($x$) on the course.

   (iv) Draw the line on your graph giving a brief description of the method you used.

   (v) Use the equation of your line to obtain a value for $y$ when $x = 10$. State in words the quantity that is being estimated by your answer.

   (vi) Explain in words precisely what the slope of your line represents in the situation under consideration.     [WJEC]

**8.** The results of an experiment to determine how the percentage sand content of soil $y$ varies with depth in cm below ground level $x$ are given in the following table.

| $x$ | 0 | 6 | 12 | 18 | 24 | 30 | 36 | 42 | 48 |
|-----|------|------|------|------|------|------|------|------|------|
| $y$ | 80.6 | 63.0 | 64.3 | 62.5 | 57.5 | 59.2 | 40.8 | 46.9 | 37.6 |

Calculate:

  (i)  the covariance of $x$ and $y$;

  (ii) the product-moment correlation coefficient of $x$ and $y$;

  (iii) the equation of the line of regression of $y$ on $x$.

Explain briefly why the product-moment correlation coefficient is preferable to the covariance as a measure of the association between $x$ and $y$.     [MEI]

**9.** In an investigation into prediction using the stars and planets, a celebrated astrologist Horace Cope predicted the ages at which thirteen young people would first marry. These complete data, of predicted and actual ages at first marriage, are now available and are summarised in the following table:

| Person | A | B | C | D | E | F | G | H | I | J | K | L | M |
|--------|----|----|----|----|----|----|----|----|----|----|----|----|----|
| **Predicted age** $x$ **(years)** | 24 | 30 | 28 | 36 | 20 | 22 | 31 | 28 | 21 | 29 | 40 | 25 | 27 |
| **Actual age** $y$ **(years)** | 23 | 31 | 28 | 35 | 20 | 25 | 45 | 30 | 22 | 27 | 40 | 27 | 26 |

  (i)  Draw a scatter diagram of these data.

  (ii) Calculate the equation of the regression line of $y$ on $x$ and draw this line on the scatter diagram.

  (iii)  Comment upon the results obtained, particularly in view of the data for person G. What further action would you suggest?     [AEB]

**10.** The following tables gives data relating to tests on ten specimens of brass.

*Exercise 4e continued*

| Specimen number | 1 | 2 | 3 | 4 | 5 | 6 | 7 | 8 | 9 | 10 |
|---|---|---|---|---|---|---|---|---|---|---|
| **Hardness $H$ in Rockwell units** | 57 | 49 | 59 | 45 | 54 | 51 | 49 | 57 | 46 | 55 |
| **Tensile strength $T$ in 1000 lbf/in²** | 76 | 69 | 83 | 64 | 74 | 73 | 66 | 79 | 65 | 80 |

Given that

$$\Sigma(H\text{--}45) = 72, \quad \Sigma(H\text{--}45)^2 = 734,$$

$$\Sigma(T\text{--}64) = 89, \quad \Sigma(T\text{--}64)^2 = 1197,$$

and $\quad \Sigma(H\text{--}45)(T\text{--}64) = 923,$

calculate the equation of the line of regression of $H$ on $T$.

A further specimen of brass has a tensile strength of 75 000 lbf/in². What do you predict about its hardness in Rockwell units? Demonstrate that the straight line method is in fact a reasonable approach to prediction in this case. [MEI]

11. The table shows the average weekly incomes in pounds sterling, after tax, for a sample of men ($x$) and women ($y$) in years 1968–1977. It also shows the incomes of men ($p$) and women ($q$) adjusted to allow for inflation.

| year | 1968 | 1969 | 1970 | 1971 | 1972 | 1973 | 1974 | 1975 | 1976 | 1977 |
|---|---|---|---|---|---|---|---|---|---|---|
| $x$ | 23.0 | 24.8 | 28.1 | 30.9 | 35.8 | 40.9 | 48.6 | 59.6 | 67.0 | 72.9 |
| $y$ | 11.3 | 12.1 | 14.0 | 15.8 | 18.3 | 21.2 | 27.0 | 34.2 | 40.6 | 44.3 |
| $p$ | 23.0 | 23.5 | 25.0 | 25.2 | 27.2 | 28.5 | 29.2 | 28.9 | 27.8 | 26.1 |
| $q$ | 11.3 | 11.5 | 12.5 | 12.9 | 13.9 | 14.8 | 16.2 | 16.6 | 16.8 | 15.9 |

$$\Sigma p = 264.4, \quad \Sigma p^2 = 7035.28, \quad \Sigma q = 142.4, \quad \Sigma q^2 = 2067.90, \quad \Sigma pq = 3802.42$$

You are given that the product moment correlation coefficient for $x$ and $y$ is 0.998. State what you would expect a scatter diagram for $x$ and $y$ to look like.

Draw a scatter diagram for $p$ and $q$, and calculate the product moment correlation coefficient for $p$ and $q$. How would you account for the difference between this correlation and that of $x$ and $y$? Which is the better indicator of the relationship between men's and women's incomes, and why?

An economist in 1977 might have used linear regression lines of income against time to predict men's and women's incomes in 1978. Explain, with reference to the figures but without doing any calculations, whether this would or would not have been a reliable procedure. [MEI]

**12.** Observations of a cactus graft were made under controlled environmental conditions. The table gives the observed heights $y$ cm of the graft at $x$ weeks after grafting. Also given are the values of $z = \ln y$.

| $x$ | 1 | 2 | 3 | 4 | 5 | 6 | 8 | 10 |
|---|---|---|---|---|---|---|---|---|
| $y$ | 2.0 | 2.4 | 2.5 | 5.1 | 6.7 | 9.4 | 18.3 | 35.1 |
| $z = \ln y$ | 0.69 | 0.88 | 0.92 | 1.63 | 1.90 | 2.24 | 2.91 | 3.56 |

(i) Draw two scatter diagrams, one for $y$ and $x$, and one for $z$ and $x$.

(ii) It is desired to estimate the height of the graft 7 weeks after grafting. Explain why your scatter diagrams suggest the use of the line of regression of $z$ on $x$ for this purpose, but not the line of regression of $y$ on $x$.

(iii) Obtain the required estimate given that $\Sigma x = 39$, $\Sigma x^2 = 255$, $\Sigma z = 14.73$, $\Sigma z^2 = 34.5231$, $\Sigma xz = 93.55$. [MEI]

**13.** The authorities in a school are concerned to ensure that their students enter appropriate Mathematics examinations. As part of a research project into this they wish to set up a performance prediction model. This involves the students taking a standard mid-year test, based on the syllabus and format of the final end-of-year examination.

The school bases its model on the belief that in the final examination students will get the same things right as they did on the mid-year test and in addition a proportion, $p$, of the things they got wrong.

Consequently a student's final mark, $y\%$ can be predicted on the basis of his or her test mark, $x\%$, by the relationship:

$$y = x + p \times (100 - x)$$

Final $=$ Test $+ p \times$ (The marks the
mark     mark            student did not
                      get on the test).

Investigate this model, using the following bivariate data. Start by finding the $y$ on $x$ regression line, and then rearrange it to estimate $p$.

| $x$ | $y$ | $x$ | $y$ | $x$ | $y$ | $x$ | $y$ |
|---|---|---|---|---|---|---|---|
| 40 | 55 | 27 | 44 | 60 | 72 | 46 | 70 |
| 22 | 40 | 32 | 50 | 50 | 70 | 70 | 85 |
| 10 | 25 | 26 | 49 | 90 | 95 | 33 | 63 |
| 46 | 68 | 68 | 76 | 30 | 50 | 40 | 60 |
| 66 | 75 | 54 | 66 | 64 | 80 | 56 | 57 |
| 8 | 32 | 68 | 70 | 100 | 100 | 45 | 55 |
| 48 | 69 | 88 | 92 | 44 | 50 | 78 | 85 |
| 58 | 66 | 48 | 59 | 58 | 62 | 68 | 80 |
| 50 | 51 | 82 | 90 | 54 | 60 | 78 | 85 |
| 80 | 85 | 66 | 76 | 24 | 31 | 89 | 91 |

*Exercise 4e continued*

**14.** A sociologist has this theory: *The proportion of people who get married, 'the marrying kind', is the same across the world. Divorce rates vary widely from one country to another, but many divorcees then remarry. Consequently a country's marriage rate depends on its divorce rate and increases in proportion to it.*

Construct a mathematical model of this theory and test it on the following data for EC countries.

| Country | Marriage Rate (/1000 population) | Divorce Rate (/1000 population) |
|---|---|---|
| Belgium | 6.5 | 2.0 |
| Denmark | 6.1 | 2.7 |
| Germany | 6.5 | 2.2 |
| Greece | 5.4 | 0.6 |
| Spain | 5.5 | 0.6 |
| France | 5.1 | 1.9 |
| Ireland | 5.0 | 0 |
| Italy | 5.4 | 0.4 |
| Luxemburg | 6.1 | 2.3 |
| Netherlands | 6.4 | 1.9 |
| Portugal | 6.9 | 0.9 |
| U.K. | 6.8 | 2.9 |

Source: *The Independent* 23/9/92

# Summary

A scatter diagram is a graph to illustrate bivariate data.

The sample covariance,

$$S_{xy} = \frac{1}{n}\Sigma(x_i - \bar{x})(y_i - \bar{y}) = \frac{1}{n}\Sigma x_i y_i - \bar{x}\bar{y}$$

The production moment correlation coefficient,

$$r = \frac{S_{xy}}{S_x S_y}$$

$$r = \frac{\Sigma(x_i - \bar{x})(y_i - \bar{y})}{\sqrt{\Sigma(x_i - \bar{x})^2 \Sigma(y_i - \bar{y})^2}}$$

An alternative form for this is

$$r = \frac{\frac{1}{n}\Sigma x_i y_i - \bar{x}\bar{y}}{\sqrt{\left(\frac{1}{n}\Sigma x_i^2 - \bar{x}^2\right)\left(\frac{1}{n}\Sigma y_i^2 - \bar{y}^2\right)}}$$

Spearman's coefficient of rank correlation

$$r_s = 1 - \frac{6\Sigma d_i^2}{n(n^2 - 1)}$$

The equation of the $y$ on $x$ regression line is

$$y - \bar{y} = s_{xy} \frac{(x - \bar{x})}{s_x^2}$$

## Exercise 4F

Collect a substantial set of raw bivariate data (at least 50 items) relating to a subject which interest you, and present it in a written report, which should include suitable graphs, calculations and interpretation.

# Appendix
# Mathematical Notes

## 1. Mean and variance of the binomial distribution

**N O T E** *This rather lengthy piece of work can be reduced to just a few lines by using probability generating functions, which you meet in Statistics 5.*

### Mean

$$E(X) = \sum_i x_i P(x = x_i) = \sum_r r P(X = r)$$

For the binomial distribution $X \sim B(n, p)$

$$P(X=r) = {}^nC_r p^r (1-p)^{n-r}$$

$$E(X) = 0 \times {}^nC_0(1-p)^n + 1 \times {}^nC_1 p(1-p)^{n-1} + 2 \times {}^nC_2 p^2 (1-p)^{n-2} + \ldots + n \, {}^nC_n p^n$$

$$= \sum_{r=0}^{n} r \, {}^nC_r p^r (1-p)^{n-r}$$

$$= \sum_{r=1}^{n} r \, {}^nC_r p^r (1-p)^{n-r} \quad \text{(since the first term is zero)}$$

$$= \sum_{r=1}^{n} r \frac{n!}{r!(n-r)!} p^r (1-p)^{n-r}$$

$$= \sum_{r=1}^{n} \frac{n!}{(r-1)!(n-r)!} p^r (1-p)^{n-r}$$

$$= np \sum_{r=1}^{n} \frac{(n-1)!}{(r-1)!(n-r)!} p^{r-1} (1-p)^{n-r}$$

$$= np \sum_{r=1}^{n} {}^{n-1}C_{r-1} p^{r-1} (1-p)^{n-r}$$

But $\quad \displaystyle\sum_{r=1}^{n} {}^{n-1}C_{r-1} p^{r-1} (1-p)^{n-r} = (p + [1-p])^{n-1} = 1$

$$\therefore E(X) = np.$$

# Variance

$$\mathrm{Var}(X) = \sum_i x_i^2\, P(X = x_i) - [E(X)]^2$$

$$= \sum_{r=0}^{n} r^2\ {}^nC_r p^r (1-p)^{n-r} - n^2 p^2$$

$$= \sum_{r=1}^{n} r^2\, \frac{n!}{r!(n-r)!} p^r (1-p)^{n-r} - n^2 p^2 \qquad \text{(since the first term is zero)}$$

$$= np \sum_{r=1}^{n} r\, \frac{(n-1)!}{(r-1)!(n-r)!}\ p^{r-1}(1-p)^{n-r} - n^2 p^2$$

$$= np \left[ \sum_{r=1}^{n} \left\{ \frac{(r-1)(n-1)!}{(r-1)!(n-r)!} + \frac{(n-1)!}{(r-1)!(n-r)!} \right\} p^{r-1}(1-p)^{n-r} \right] - n^2 p^2$$

$$= np \left[ \sum_{r=2}^{n} (r-1)\, \frac{(n-1)!}{(r-1)!(n-r)!} p^{r-1}(1-p)^{n-r} \right.$$

$$\left. + \sum_{r=1}^{n} \frac{(n-1)!}{(r-1)!(n-r)!} p^{r-1}(1-p)^{n-r} \right] - n^2 p^2$$

$$= np \left[ (n-1)p \sum_{r=2}^{n} \frac{(n-2)!}{(r-2)!(n-r)!} p^{r-2}(1-p)^{n-r} + \left\{ p + (1-p) \right\}^{n-1} \right] - n^2 p^2$$

$$= np \left[ (n-1)p \left\{ [p+(1-p)]^{n-2} \right\} + 1 \right] - n^2 p^2$$

$$= np[(n-1)p + 1] - n^2 p^2$$

$$= n^2 p^2 - np^2 + np - n^2 p^2$$

$$= np(1-p)$$

$$= npq$$

# 2. Mean and Variance of the Poisson distribution

## Mean

$$E(X) = \sum_i x_i P(X = x_i) = \sum_{r=0}^{\infty} r P(X = r)$$

$$= 0 + 1 \times P(X = 1) + 2 \times P(X = 2) + 3 \times P(X = 3) + \cdots$$

$$= \frac{1.\lambda e^{-\lambda}}{1!} + \frac{2.\lambda^2 e^{-\lambda}}{2!} + \frac{3.\lambda^3 e^{-\lambda}}{3!} + \cdots$$

$$= \lambda e^{-\lambda}\left(1 + \lambda + \frac{\lambda^2}{2!} + \frac{\lambda^3}{3!} + \cdots\right)$$

But the series in the brackets is just $e^{\lambda}$

$$\therefore E(X) = \lambda e^{-\lambda}e^{\lambda}$$

$$= \lambda$$

## Variance

$$\text{Var}(X) = E(X^2) - [E(X)]^2$$

$$E(X^2) = \sum_{r=0}^{\infty} r^2 P(X = r)$$

Putting

$$r^2 = r(r-1) + r$$

gives

$$E(X^2) = \sum_{r=0}^{\infty}\{r(r-1) + r\}P(X = r)$$

$$= \sum_{r=0}^{\infty} r(r-1)P(X = r) + \sum_{r=0}^{\infty} rP(X = r)$$

Now

$$\sum_{r=0}^{\infty} rP(X = r) = \lambda$$

$$\therefore \quad E(r^2) = \sum_{r=0}^{\infty} r(r-1)P(X = r) + \lambda$$

Now the first term

$$\sum_{r=0}^{\infty} r(r-1)P(X = r) = 0 + 0 + 2\times1\times P(X = 2) + 3\times2\times P(X = 3) + 4\times3\times P(X = 4) + \cdots$$

$$= \frac{2\times1\times\lambda^2 e^{-\lambda}}{2!} + \frac{3\times2\times\lambda^3 e^{-\lambda}}{3!} + \frac{4\times3\times\lambda^4 e^{-\lambda}}{4!} + \cdots$$

cancelling

$$= \lambda^2 e^{-\lambda} + \lambda^3 e^{-\lambda} + \frac{\lambda^4 e^{-\lambda}}{2!} + \frac{\lambda^5 e^{-\lambda}}{3!} + \cdots$$

$$= \lambda^2 e^{-\lambda}\left(1 + \lambda + \frac{\lambda^2}{2!} + \frac{\lambda^3}{3!} + \cdots\right)$$

Once again the term in brackets is $e^\lambda$

$$
\begin{aligned}
&= \lambda^2 e^{-\lambda} e^\lambda \\
&= \lambda^2
\end{aligned}
$$

This now gives:

$$E(X^2) = \lambda^2 + \lambda$$

So that:

$$
\begin{aligned}
\text{Var}(X) &= \lambda^2 + \lambda - (\lambda)^2 \\
&= \lambda
\end{aligned}
$$

Thus showing that for the Poisson distribution the mean and variance are both $\lambda$.

# 3. The Sum of two independent Poisson distributions

Suppose that $X \sim$ Poisson $(\lambda)$ and $Y \sim$ Poisson $(\mu)$ and that $X$ and $Y$ are independent, and that $Z$ is the random variable given by

$$Z = X + Y$$

$P(Z=z) = P(X=z) \times P\,(Y=0) + P\,(X=z{-}1) \times P\,(Y=1) + P(X=z{-}2) \times P\,(Y=2) + \ldots + P\,(X=0) \times P\,(Y=z)$

$X$ and $Y$ must be independent so that the probabilities can be multiplied.

$$
= \frac{\lambda^z e^{-\lambda} e^{-\mu}}{z!} + \frac{\lambda^{z-1} e^{-\lambda} \mu e^{-\mu}}{(z-1)!1!} + \frac{\lambda^{z-2} e^{-\lambda} \mu^2 e^{-\mu}}{(z-2)!2!} + \cdots + \frac{e^{-\lambda} \mu^z e^{-\mu}}{z!}
$$

$$
= e^{-(\lambda+\mu)} \left\{ \frac{\lambda^z}{z!} + \frac{\lambda^{z-1}\mu}{(z-1)!1!} + \frac{\lambda^{z-2}\mu^2}{(z-1)!2!} + \cdots + \frac{\mu^z}{z!} \right\}
$$

$$
= \frac{e^{-(\lambda+\mu)}}{z!} \left\{ \lambda^z + \frac{z!\lambda^{z-1}\mu}{(z-1)!1!} + \frac{z!\lambda^{z-2}\mu^2}{(z-1)!2!} + \cdots + \mu^z \right\}
$$

Now the term in bracket is just the binomial expansion of $(\lambda + \mu)^z$

$$P(Z = z) = \frac{e^{-(\lambda+\mu)}(\lambda+\mu)z}{z!}$$

which is the Poisson probability with parameter $\lambda + \mu$

So the sum of the two independent Poisson distributions with parameters $\lambda$ and $\mu$ is itself a Poisson distribution with parameter $\lambda + \mu$.

Similarly the sum of $n$ independent Poisson distributions with parameters $\lambda_1, \lambda_2, ..., \lambda_n$ is a Poisson distribution with parameter $(\lambda_1 + \lambda_2 + ... + \lambda_n)$.

# 4. Equivalence of Spearman's rank correlation coefficient and Pearson's product moment correlation coefficient

Let the ranks assigned to $n$ items by two judges be $x_1, x_2, ..., x_n$ and $y_1, y_2, ..., y_n$

$$x_1 \qquad x_2 \qquad \cdots x_i \qquad \cdots x_n$$

$$y_1 \qquad y_2 \qquad \cdots y_i \qquad \cdots x_n$$

$$|d_i| = |x_1 - y_1| \quad |x_2 - y_2| \quad \cdots \quad |x_i - y_i| \quad \cdots \quad |x_n - y_n|$$

$$\sum_{i=1}^{n} d_i^2 = \sum_{i=1}^{n} (x_i - y_i)^2 = \sum x_i^2 + \sum y_i^2 - 2\sum_i x_i y_i$$

Since the values of $x_1 ... x_n$ and $y_1 ... y_n$ are both 1, 2, 3, ..., $n$ in some order

$$\sum_{i=1}^{n} x_i = \sum_{i=1}^{n} y_i = \frac{n(n+1)}{2} \qquad \bar{x} = \bar{y} = \frac{n+1}{2}$$

$$\sum_{i=1}^{n} x_i^2 = \sum_{i=1}^{n} y_i^2 = \frac{n(n+1)(2n+1)}{6} \qquad s_x = s_y = \sqrt{\frac{n(n+1)(2n+1)}{6n} - \left(\frac{n+1}{2}\right)^2}$$

$$= \sqrt{\frac{(n+1)(n-1)}{12}} = \sqrt{\frac{n^2 - 1}{12}}$$

$$r_s = 1 - \frac{6\Sigma d_i^2}{n(n^2 - 1)}$$

$$= 1 - \frac{6[\Sigma x_i^2 + \Sigma y_i^2 - 2\Sigma x_i y_i]}{n(n^2 - 1)}$$

$$= \frac{1}{n(n^2 - 1)}\left\{ n(n^2 - 1) - 6\left[ \frac{n(n+1)(2n+1)}{6} + \frac{n(n+1)(2n+1)}{6} - 2\Sigma x_i y_i \right]\right\}$$

$$= \frac{12\Sigma x_i y_i - 3n(n+1)^2}{n(n^2 - 1)}$$

$$= \frac{12\left[\dfrac{\Sigma x_i y_i}{n} - \left(\dfrac{n+1}{2}\right)^2\right]}{(n^2-1)}$$

$$= \frac{\displaystyle\sum \frac{x_i y_i}{n} - \bar{x}, \bar{y}}{\sqrt{\dfrac{n^2-1}{12}}\sqrt{\dfrac{n^2-1}{12}}}$$

$$= \frac{S_{xy}}{S_x S_y}$$

$$= r \qquad \text{as required}$$

# 5. Least Squares Regression Line

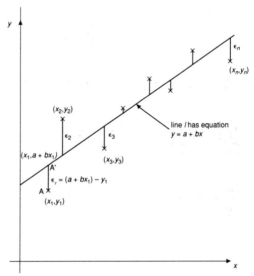

The least squares regression line for the set of bivariate data $(x_1, y_1)$, $(x_2, y_2)$, ..., $(x_n, y_n)$ is of the form

$$y = a + bx$$

with the values of $a$ and $b$ giving the minimum sum of the squares of the residuals, $\varepsilon_1, \varepsilon_2, \ldots, \varepsilon_n$.

Since $\varepsilon_r = a + bx_r - y_r$, it is required to find the values of $a$ and $b$ which minimise

$$S = \sum_{r=1}^{n} (a + bx_r - y_r)^2$$

Since $S$ has to be a minimum with respect to both $a$ and $b$, it is necessary, in turn, to hold one of $a$ or $b$ constant and then differentiate $S$ with respect to the other, a technique known as partial differentiation; the symbol d is replaced by $\partial$ to show this.

Holding $b$ constant

$$\frac{\partial S}{\partial a} = 2\sum_{r=1}^{n}(a + bx_r - y_r)$$

$$= 0 \text{ for } S \text{ to be a minimum}$$

$$\therefore \quad \sum_{r=1}^{n}a + b\sum_{r=1}^{n}x_r - \sum_{r=1}^{n}y_r = 0$$

and since

$$\bar{x} = \frac{1}{n}\sum_{r=1}^{n}x_r, \bar{y} = \frac{1}{2}\sum_{r=1}^{n}y_r \text{ and } \sum_{r=1}^{n}a = na$$

this gives

$$na + bn\bar{x} - n\bar{y} = 0$$
$$a + b\bar{x} = \bar{y} \tag{1}$$

Notice that this means the mean point $(\bar{x}, \bar{y})$ lies on $y = a + bx$

Holding $a$ constant

$$\frac{\partial S}{\partial b} = \sum_{r=1}^{n}(a + bx_r - y_r)x_r$$

$$= 0 \text{ for } S \text{ to be a minimum}$$

$$\therefore \quad a\sum_{r=1}^{n}x_r + b\sum_{r=1}^{n}x_r^2 - \sum_{r=1}^{n}x_r y_r = 0$$

Dividing through by $n$ gives

$$a\bar{x} + b\frac{1}{n}\sum_{r=1}^{n}x_r^2 = \frac{1}{n}\sum_{r=1}^{n}x_r y_r \tag{2}$$

Equations (1) and (2) are simultaneous equations for $a$ and $b$.

$$(2) \qquad a\bar{x} + b\frac{1}{n}\sum_{r=1}^{n}x_r^2 = \frac{1}{n}\sum_{r=1}^{n}x_r y_r$$

$$(1) \times \bar{x} \qquad \underline{\qquad a\bar{x} + b\bar{x}^2 \qquad = \bar{x}\bar{y} \qquad}$$

Subtract

$$b\left[\frac{1}{n}\sum_{r=1}^{n}x_r^2 - \bar{x}^2\right] = \frac{1}{n}\sum_{r=1}^{n}x_r y_r - \bar{x}\bar{y}$$

but this is

$$bs_x^2 = s_{xy}$$

therefore

$$b = \frac{s_{xy}}{s_x^2}$$

Substituting for $b$ in equation (1) gives

$$a = \bar{y} - \frac{S_{xy}}{S_x^{\,2}}\bar{x}$$

So $y = a + bx$ becomes

$$y = \bar{y} - \frac{S_{xy}}{S_x^{\,2}}\bar{x} + \frac{S_{xy}}{S_x^{\,2}}x$$

or

$$y - \bar{y} = \frac{S_{xy}}{S_x^{\,2}}(x - \bar{x})$$

and this is the equation of the least squares regression line.

# Answers to selected questions

## 1. DISCRETE RANDOM VARIABLES
### Exercise 1A

1. $k = \frac{1}{10}$
2. Number of turns needed to obtain a '6' on a die.
3. (i) $a = 0.4$  (ii) 0.3
4. 

| X | 0 | 1 | 2 | 3 | 4 | 5 |
|---|---|---|---|---|---|---|
| Probability | $\frac{1}{32}$ | $\frac{5}{32}$ | $\frac{10}{32}$ | $\frac{10}{32}$ | $\frac{5}{32}$ | $\frac{1}{32}$ |

5. $c = \frac{1}{3}$, 0.2
6. 

| X | 2 | 3 | 4 | 5 | 6 | 7 | 8 | 9 | 10 | 11 | 12 |
|---|---|---|---|---|---|---|---|---|----|----|----|
| Probability | $\frac{1}{36}$ | $\frac{2}{36}$ | $\frac{3}{36}$ | $\frac{4}{36}$ | $\frac{5}{36}$ | $\frac{6}{36}$ | $\frac{5}{36}$ | $\frac{4}{36}$ | $\frac{3}{36}$ | $\frac{2}{36}$ | $\frac{1}{36}$ |

7. (i) 

| Y | 0 | 1 | 2 | 3 | 4 | 5 |
|---|---|---|---|---|---|---|
| Probability | $\frac{6}{36}$ | $\frac{10}{36}$ | $\frac{8}{36}$ | $\frac{6}{36}$ | $\frac{4}{36}$ | $\frac{2}{36}$ |

(ii) $\frac{2}{3}$

8. (i) 

| Z | 0 | 1 | 2 | 3 | 4 |
|---|---|---|---|---|---|
| Probability | $\frac{1}{16}$ | $\frac{4}{16}$ | $\frac{6}{16}$ | $\frac{4}{16}$ | $\frac{1}{16}$ |

(ii) $\frac{5}{16}$

9. 

| X | 0 | 1 | 2 | 3 |
|---|---|---|---|---|
| Probability | 0.1667 | 0.5 | 0.3 | 0.0333 |

10. 

| Number of men | 0 | 1 | 2 | 3 |
|---|---|---|---|---|
| Probability | 0.122 | 0.441 | 0.367 | 0.070 |

11. (i) 

| X | 1 | 2 | 3 | 4 | 6 | 8 | 9 | 12 | 16 |
|---|---|---|---|---|---|---|---|----|----|
| Probability | $\frac{1}{16}$ | $\frac{2}{16}$ | $\frac{2}{16}$ | $\frac{3}{16}$ | $\frac{2}{16}$ | $\frac{2}{16}$ | $\frac{1}{16}$ | $\frac{2}{16}$ | $\frac{1}{16}$ |

(ii) 0.25

12. 

| Number of red cards | 0 | 1 | 2 | 3 | 4 |
|---|---|---|---|---|---|
| Probability | 0.055 | 0.25 | 0.39 | 0.25 | 0.055 |

13. (i) $k = 0.08$

| X | 0 | 1 | 2 | 3 |
|---|---|---|---|---|
| Prob | 0.2 | 0.24 | 0.32 | 0.24 |

(ii) 

| Number of chicks surviving | 0 | 1 | 2 | 3 |
|---|---|---|---|---|
| Probability | 0.35104 | 0.44928 | 0.18432 | 0.01536 |

14. (i) $a = 0.42$  (ii) $k = \frac{1}{35}$  (iii) Algebraic model is not particularly accurate. No.
15. (i) $\frac{1}{1296}$  (ii) $\frac{1}{81}$  (iii) $\frac{65}{1296}$  (iv) $\frac{671}{1296}, > \frac{1}{2}$

### Exercise 1B

1. 1.5
2. 2.7
3. $P(X = 4) = 0.8$, $P(X = 5) = 0.2$
4. 

| Y | 50 | 100 |
|---|---|---|
| Probability | 0.4 | 0.6 |

5. 3.22
6. 60 pence
7. (i) loss of 2.5p  (ii) loss of 7.5p  (iii) loss of £2.50
8. (i) £168
9. (i) 2.79  (ii) 8.97  (iii) 20.94
10. (i) $w = 21$  (ii) $w = 27.67$
11. (ii) 0.13  (iii) 0.73  (iv) $E(X) = 24$
12. (i) 3 coins  (ii) £1.30
13. (i) 0.0405  (ii) 0.1118

14. $^4/_{35}$, $^{18}/_{35}$, $^{12}/_{35}$, $^1/_{35}$, $^{£(54+r)}/_{35}$, $r = 50$
15. (i) (A) $^{32}/_{80}$ (B) $^{36}/_{48}$ (C) $^3/_{12}$ (ii) $^9/_{80}$ (iii) $^{36}/_{71}$ (iv) £563.38
    (v) One week, £625: two week, £494.12. Two week course.

## Exercise 1C

1. (i) $E(X) = 3.1$ (ii) $Var(X) = 1.29$
2. (i) $E(X) = 0.7$ (ii) $Var(X) = 0.61$
4. (i) $E(2X) = 6$ (ii) $Var(3X) = 6.75$
5. (i) $k = 0.1$ (ii) 1.25 eggs (iii) 0.942
6. (i) $E(Y) = 1.2857$ (ii) $Var(Y) = 0.49$
7. (i) 10.9, 3.09; (ii) 18.4, 111.24
8. 4.333, 2.22, 8.666, 8.88.
9. (i) 2 (ii) 1 (iii) 9
10. £5.47, 0.086
11. (i) $p = ^1/_{16}$ (ii) $E(X) = 2$, $Var(X) = 3.5$ (iii) $E(2X + 3) = 7$, $Var(2X + 3) = 14$
    (iv) 4; 0.281
12. 

| $X$ | 6 | 7 | 8 | 9 | 10 | S.D. = 0.975; 0.6397 |
|---|---|---|---|---|---|---|
| **Probability** | $^6/_{72}$ | $^{24}/_{72}$ | $^{24}/_{72}$ | $^{16}/_{72}$ | $^2/_{72}$ | |

13. 2.449, 2.57
14. 

| $N$ | 2 | 3 | 4 | 5 | 6 | 7 | 8 | $E(N)=5$, $Var(N)=2.5$ |
|---|---|---|---|---|---|---|---|---|
| **Probability** | $^1/_{16}$ | $^2/_{16}$ | $^3/_{16}$ | $^4/_{16}$ | $^3/_{16}$ | $^2/_{16}$ | $^1/_{16}$ | |

15. 

| $X$ | $-1$ | $^{-2}/_3$ | $^{-1}/_3$ | 0 | $^1/_3$ | $^2/_3$ | 1 | $Var(X)=^1/_6$ |
|---|---|---|---|---|---|---|---|---|
| **Probability** | $^1/_{64}$ | $^6/_{64}$ | $^{15}/_{64}$ | $^{20}/_{64}$ | $^{15}/_{64}$ | $^6/_{64}$ | $^1/_{64}$ | |

16. (ii) 3.5, 0.897 (iii) 0.5346 (iv) 0.932
17. (i) 

| $X$ | 0 | 1 | 2 | 3 | 4 | 5 | 6 | 7 | 8 |
|---|---|---|---|---|---|---|---|---|---|
| **Probability** | 0.033 | 0.1 | 0.133 | 0.167 | 0.167 | 0.167 | 0.122 | 0.089 | 0.022 |

    (ii) 3.9, 3.93 (iii) $k = ^1/_{84}$ (iv) 4, 3 (v) Yes.
18. (iii) 2.67, 2.22 (v) (A): 1.62, 0.52 (B): 1.56, 0.68

## Exercise 1D

1. (i) 

| $X$ | 0 | 1 | 2 | 3 | 4 | 5 | 6 | 7 | 8 | 9 | 10 |
|---|---|---|---|---|---|---|---|---|---|---|---|
| **Probability** | $\frac{1}{1024}$ | $\frac{10}{1024}$ | $\frac{45}{1024}$ | $\frac{120}{1024}$ | $\frac{210}{1024}$ | $\frac{252}{1024}$ | $\frac{210}{1024}$ | $\frac{120}{1024}$ | $\frac{45}{1024}$ | $\frac{10}{1024}$ | $\frac{1}{1024}$ |

    (ii) 5, 2.5
2. (i) 

| $Y$ | 0 | 1 | 2 | 3 | 4 | 5 |
|---|---|---|---|---|---|---|
| **Probability** | 0.4019 | 0.4019 | 0.1608 | 0.0322 | 0.0032 | 0.0001 |

    (ii) 0.83333, 0.6944
3. (i) 5, 2.5 (ii) 5.15, 1.8275 (iii) Yes
4. (i) 

| $Y$ | 5 | 10 | 15 | 20 | 25 | 30 | (ii) £17.50, £8.54 |
|---|---|---|---|---|---|---|---|
| **Prob** | $^1/_6$ | $^1/_6$ | $^1/_6$ | $^1/_6$ | $^1/_6$ | $^1/_6$ | |

    (iii) Perhaps someone has denoted an additional £1000!
5. (i) 2 (ii) 3 (iii) First couple: $^1/_8$; second couple: $^1/_4$
6. (i) Geometric distribution with $p = 0.4$. (ii) 2.5, 3.75
7. (i) 

| $X$ | 1 | 2 | 3 | 4 | 2.5, 1.25 |
|---|---|---|---|---|---|
| **Prob** | $^1/_4$ | $^1/_4$ | $^1/_4$ | $^1/_4$ | |

    (ii) 

| $Y$ | 2 | 3 | 4 | 5 | 6 | 7 | 8 | 5, 2.5 (iii) 3 packets |
|---|---|---|---|---|---|---|---|---|
| | $^1/_{16}$ | $^2/_{16}$ | $^3/_{16}$ | $^4/_{16}$ | $^3/_{16}$ | $^2/_{16}$ | $^1/_{16}$ | |

    (iv) 0.859
8. 1.66, $p = 0.553$, 0.982

# 2. THE POISSON DISTRIBUTION

## Exercise 2A

1.  (i)  0.271   (ii) 0.090
2.  (i)  0.175   (ii) 0.104
3.  (i)  0.082   (ii) 0.214   (iii) 0.067
4.  (i)  0.050   (ii) 0.149   (iii) 0.224   (iv) 0.423   (v) 0.577
5.  (i)  0.359   (ii) 0.641
6.  (i)  0.1912   (ii) 0.8088
7.  (i)  0.165   (ii) 0.835
8.  (i)  25   (ii) 75   (iii) 112   (iv) 278
9.  (i)  42   (ii) $\bar{x}=2$ $s^2 = 2.381$;   (iii) 0.135, 0.271, 0.271, 0.180, 0.090, 0.036, 0.012
    (iv) 5.7, 11.4, 11.4, 7.6, 3.8, 1.5, 0.5
10. (i)  0.111   (ii) 0.244   (iii) 0.268   (iv) 0.377   (v) 3
11. (i)  0.135   (ii) 0.271   (iii) 36 tubs,  2.6 complaints
12. (i)  0.738   (ii) 239, 177, 65, 16, 0, 0   (iii) 239.0, 176.4, 65.1, 16.0, 3.0, 0.4
13. (i)  0.3328   (ii) .0016
14. 0.082;  0.109;  0.779;  0.207
15. $\bar{x} = 0.5$,  $s^2 = 0.49$;  31.5,  15.8,  3.9,  0.7,  0.1
16. (i)  0.030   (ii) 0.275,  461
17. $\bar{x} = 7.03$,  6.84;  8.17
18. (i)  0.165   (ii) 5   (iii) 0.027   (iv) 5
19. (iii) 0.012

## Exercise 2B

1.  (i)  0.362   (ii) 0.544   (iii) 0.214   (iv) 0.558
2.  (i)  0.082   (ii) 0.891;  0.287
3.  (i)  0.161   (ii) 0.554   (iii) 10   (iv) 0.016   (v) 0.119
4.  (i)  0.101   (ii) 0.285   (iii) 0.422
5.  (i)  0.175   (ii) 0.973   (iii) 0.031;  0.125;  0.249
6.  (i)  0.175   (ii) 0.560   (iii) 0.102   (iv) 0.534   (v) 0.005, 10
7.  13.9%
8.  0.014;  0.205
9.  (i)  0.135   (ii) 0.9473   (iii) 0.0527   (iv) 13   (v) 0.1242
10. (i)  0.082   (ii) 0.456   (A) 0.309   (B) 1 eastbound & 2 westbound
11. $\bar{x} = 3.87$; $s^2 = 3.53$.   Using $\lambda = 3.87$, frequencies 10.8, 41.8, 81.0, 104.5, 101.2, 78.4, 50.6,
    28.0, 13.5, 9.2

# 3 THE NORMAL DISTRIBUTION

## Exercise 3A

1.  (i)  0.8413   (ii) 0.0228   (iii) 0.1359
2.  (i)  0.0668   (ii) 0.6915   (iii) 0.2417
3.  (i)  0.0668   (ii) 0.1587   (iii) 0.7745
4.  (i)  0.0038   (ii) 0.5   (iii) 0.495
5.  (i)  31, 1.97   (ii) 2.1, 13.4, 34.5, 34.5, 13.4, 2.1
6.  63g
7.  (i)  5.48%   (ii) (a) 25425 km   (b) 1843 km
8.  (i)  78.65%   (ii) 5.254,  0.054
9.  (i)  11.37%   (ii) 0.4996   (iii) 0.17
10. (i)  0.0774   (ii) 0.2216  0.4315
11. (i)  0.0766   (ii) 0.8468   (iii)      0.6743  1.313 m
12. (i)  0.0900   (ii) 0.5392   (iii) 0.3467, 8.28' 30"
13. 0.0401   (i) 0.4593   (ii) 0.003
14. (i)  0.675   (ii) 0.325   0.481, 31 days
15. 20.0491, 0.0248, 0.7794, 22.6%

16. 74.35, 3.23, (i) 0.756 (ii) 0.244
17. 0.0312 (i) 0.937 (ii) 0.08
18. (i) B (ii) A A(£15.16) B(£10.96)
19. 11.09%, 0.0123, 0.0243
20. (i) 14.8 (ii) 0.0668 (iii) 1.56

## Exercise 3B

1. 0.165
2. 106, 75 and 125, 39.5
3. 0.282, 0.526, 25, 4.33, 0.102
4. 0.246, 0.0796, 0.0179
5. 100, 9.49, 122.5
6. (a) .315 (b) .291, Worse .5245
7. $1/3$, $6^2/3$, $^{40}/9$, 13
8. (i) Almost 1 (ii) 0.98
9. Almost zero, 0.0044, 73

10. $^{2000}C_N \left( \dfrac{1}{30} \right)^N \left( \dfrac{29}{30} \right)^{2000-N}$   86   More (96)

11. 0.180, 60, 7.75, 0.9125
12. 0.0222, 0.9778
13. 0.887 0.994, 18, 0.277
14. (i) 0.135 (ii) 0.271 (iii) 0.271 (iv) 2.996 (v) 0.175 (vi) 0.804
15. (i) 2.5 (ii) 0.918, 0.358 (iii) 0.158

# 4 BIVARIATE DATA
## Exercise 4B

1. $S_{xy} = -7$, $r = -0.96$
2. $S_{xy} = 11$, $r = 0.704$
3. $S_{xy} = -3.2$, $r = -0.924$
4. $S_{xy} = -8.875$, $r = -0.635$
5. $S_{xy} = -54.5$, $r = -0.128$

## Exercise 4C

1. (i) 0.913 (ii) $H_0: \rho = 0, H_1: \rho > 0$ (iii) Accept $H_1$.
2. (i) 0.380 (ii) $H_0: \rho = 0, H_1: \rho > 0$ (iii) Accept $H_0$.
3. (i) 0.715 (ii) $H_0: \rho = 0, H_1: \rho > 0$ (iii) Accept $H_1$.
4. (i) 0.850 (ii) $H_0: \rho = 0, H_1: \rho > 0$ (iii) Accept $H_1$.
5. (i) 0.946 (ii) $H_0: \rho = 0, H_1: \rho > 0$ (iii) Accept $H_1$.
6. (i) 0.901 (ii) $H_0: \rho = 0, H_1: \rho > 0$ (iii) Accept $H_1$.
7. (i) $H_0: \rho = 0$, $H_1: \rho \neq 0$, 5% sig. level (ii) 0.491, accept $H_1$. (iii) (18.8. 45), (18.2, 45), (18.7, 45), it seems as though these girls stopped for a rest.
8. (i) 0.807 (ii) $H_0: \rho = 0$, $H_1: \rho > 0$ (iii) Accept $H_1$.
9. (i) $H_0: \rho = 0$, $H_1: \rho < 0$. $r = -0.854$, accept $H_1$.
10. $r = 0.59$. Diagram suggests moderate positive correlation which is confirmed by the fairly high positive value of $r$ . $r = -0.145$. Discarding high and low values of $x$ seem to produce an uncorrelated set.
11. (i) 0.032 (ii) −0.862 (iii) 0.0648 (iv) 0.988. Cases (i) and (iii) indicate no correlation, (ii) strong negative correlation, and (iv) strong positive correlation, indicating a 4 quarter cyclic variation in the data.

## Exercise 4D

1. 0.143
2. 0.576
3. 0.8576

4.  (i) –0.479   (ii) $H_0: \rho = 0$, $H_1: \rho > 0$, judges not in agreement.
5.  –0.0875
6.  (i)  Y and Z   (ii) $H_0: \rho = 0$, $H_1: \rho > 0$, accept $H_1$ at 5% significance level
7.  –0.821. Judges disagree with each other.
8.  (i)  0.766, –0.143   (ii) $H_0: \rho = 0$, $H_1: \rho > 0$, accept $H_1$.
    (iii) Product-moment correlation coefficient is more suitable here because it takes into account the magnitude of the variables.
9.  –0.690. $H_0: \rho = 0$, $H_1: \rho < 0$, accept $H_1$.
10. (i)  0.636. Positive sign indicates possible positive linear correlation.
    (ii)  $H_0: \rho = 0$, $H_1: \rho > 0$, accept $H_1$.
11. (i)  0.680   (ii) $H_0: \rho = 0$, $H_1: \rho > 0$, 0.680 > 0.6694 so accept $H_1$.
    (iii) 0.214   (iv) Mr. Smith ought to have used Spearman's coefficient.
12. (i)  0.881  (ii) 0.888  (iii) 0.622  significant positive correlation at 1% level in all cases
13. (i)  0.636   (ii) $H_0: \rho = 0$, $H_1: \rho > 0$, accept $H_0$   (iii) It is more appropriate to use the product moment correlation coefficient which utilises the actual data values.

# Exercise 4E

1.  $19x + 50y = 1615$, 27.7
2.  $1.446x + y = 114.38$, 53.7
3.  (i)  $y = 8.45 + 27.5$
    (iii) Value of these stamps is not compatible with the earlier part of the question.
4.  (ii)  $0.0289x + y = 0.925$   (iii) 0.434
    (iv) Stem density is outside domain of validity. If $x > 32$ model predicts that $y$ is negative!
5.  $1.013x + y = 64.25$, 44%. Second treatment seems to produce fewer grubs but linear correlation is not as strong as before.
6.  $\bar{x} = 4.62$, $\bar{y} = 157.3$, cancer mortality = 179.3
7.  (i)  $r = 0.984$, it is significantly different from zero.   (iii) $y = 1.019x - 2.728$
    (v)  $y = 7.46$. Weight loss in kg is being estimated.   (vi) Ratio of covariance to variance of $x$ data.
8.  (i)  –180.4   (ii)  $r = -0.926$  (iii) $0.752x + y = 74.97$. Product-moment correlation coefficient is preferable because it is a standardised measure of correlation whereas covariance is not.
9.  (ii)  $y = 1.031x + 0.528$   (iii) Predictions seem very accurate despite the data for person G. A formal hypothesis test is the next step.
10. $H = 0.697T + 1.391$; 53.7
11. A straight line almost. $r = 0.884$ The correlation coefficient using $p$ and $q$ gives a better indicator. The 1978 prediction is not reliable because it extends outside the domain of validity.
12. (iii) 12.8
13. Regression line of $y$ on $x$: $y = 0.78x + 24.0$ or approx $y = x + \frac{1}{4}(100 - x)$ Thus $p = \frac{1}{4}$.
14. $m = 5.34 + 0.41\,d$.

**The University of London Examinations and Assessment Council, The University of Cambridge Local Examinations Syndicate and The Associated Examining Board accept no responsibility what so ever for the accuracy or method of working in the answers given.**

# Index